Profitable Organic

D1388726

3

631.584 NEW

Profitable Organic Farming

Second Edition

Jon Newton
MA (Oxon), MSc (Wales)
Consultant in Organic Agriculture

Blackwell
Science

© 1995, 2004 by Blackwell Science Ltd,
a Blackwell Publishing Company

Editorial offices:
Blackwell Science Ltd, 9600 Garsington
Road, Oxford OX4 2DQ, UK
Tel: +44 (0) 1865 776868
Blackwell Publishing Professional, 2121 State
Avenue, Ames, Iowa 50014-8300, USA
Tel: +1 515 292 0140
Blackwell Science Asia Pty Ltd, 550 Swanston
Street, Carlton, Victoria 3053, Australia
Tel: +61 (0)3 8359 1011

First Edition published 1995 by Blackwell
Science Ltd
Second Edition published 2004

Library of Congress Cataloging-in-
Publication Data
Newton, Jon.
    Profitable organic farming/Jon Newton.—
2nd ed.
        p. cm.
    Includes bibliographical references and
index.
    ISBN 0-632-05959-1 (alk. paper)
        1. Organic farming.   I. Title.

S605.5N48   2004
631.5′84′068—dc21   2003044335

ISBN 0-632-05959-1

A catalogue record for this title is available
from the British Library

Set in 11/13pt Times
by DP Photosetting, Aylesbury, Bucks
Printed and bound in India
by Replika Press Pvt. Ltd.

The publisher's policy is to use permanent
paper from mills that operate a sustainable
forestry policy, and which has been
manufactured from pulp processed using acid-
free and elementary chlorine-free practices.
Furthermore, the publisher ensures that the
text paper and cover board used have met
acceptable environmental accreditation
standards.

For further information on Blackwell
Publishing, visit our website:
www.blackwellpublishing.com

For Mary, Seán and Cecilia

Contents

Foreword to the First Edition

I first learned about farming from my father who went to Aspatria College in Cumbria in 1908. I saw that growing crops and grass without chemicals was possible, and did work, and when I took over this 400 ha farm in 1976 I wished to put those ideas into practice. However, I was bombarded with advice from clever young experts. It was the mid-seventies and the height of the barley boom. They told me I must tighten up the stock, use a great deal of nitrogen, and plough up every possible field for cropping. Fortunately, basic instinct told me that, especially on an upland farm, this would be the road to ruin, and so it has proved to be for many a farmer.

So we went the other way and instead of spending money on nitrogen and machinery we spent it on insoluble phosphates, lime, clover, and seaweed products, both liquid and mineral. We are now able to carry more breeding stock than before, and to finish all our lambs and almost all our calves ourselves. Our vet bills are very much lower than before. Had I gone the other way, I feel sure I would now have impoverished land which would be very difficult to get back to high grass production carrying a lot of stock. I believe that many more people would become organic if they felt that they would receive the same level of support as our French and German neighbours do. Organic farmers are, after all, doing exactly what the Common Agricultural Policy (CAP) reformers require, with no over-production and a healthier environment. We do not pollute the rivers, tear up hedges, nor poison food. Our animals enjoy a high level of welfare. Surely, in the end, this will be realised and perhaps recognised.

This book by Jon Newton shows us the way to successful organic farming, and provides targets showing what is possible in an organic situation, a situation which is not always easy. For example, there is no readily available source of organically approved potash. It would be convenient if we were permitted to put a very small amount of calcium nitrate on our silage fields. It can be difficult to grow fodder rape. But these are difficulties which can be overcome, whereas if we had been farming conventionally and then found our fields had become less productive, or that the health of our stock was breaking down perhaps due to mineral deficiencies, it would be more difficult still, and take time to correct.

As far as the finances of organic farming are concerned, I am far from convinced that I could make any more money in a conventional system. Yes, we could grow more silage and barley, and finish more bullocks. But to increase breeding stock would need a massive investment in sheds. Expenditure would rise on nitrates and fertilisers. Vet bills would most certainly rise. The economic arguments are far from simple, but how easily one can be misled by persistent salesmen! Managing organic units needs people who are personally committed in their beliefs, and able to inspire others with similar enthusiasm. Not for them the easy solutions of the penicillin bottle for every illness, nor the quick spray with weed killer for a different crop. They must foresee their problems and take preventative measures.

As to the future of organics, I remain optimistic. The barley plain is in retreat, and one hopes that some farmers will revert to a more sustainable mode of agriculture. It is here that Jon Newton's book will prove invaluable, setting technical standards and advising on finance, marketing and sales.

The late Anne Scott
Galashiels
May 1994

Acknowledgements

I would like to thank the following organic farmers who willingly found time to discuss their farms with Chan Sook Cheng and who have given me a great deal of encouragement: C.C. Adams, Henry and Sally Bagenal, Karl and Christine Barton, Will and Pam Best, Oliver Dowding, John Eliot Gardiner, Ed and Sheila Goff, Mark Houghton-Brown, Jo Phillips, the late Anne Scott, Jamie Turnbull, the late Willie Warnock, Elaine Whewell and David Wilson.

Chan Sook Cheng contributed Chapters 8 and 9, and has added a dimension to the book which I hope has much extended its usefulness. Will Best contributed Chapter 11, from the standpoint of a practising farmer. I am also most grateful to Ed Goff for his comments and helpful criticism of the dairy section (Chapter 3) and Chapter 5. Helen Browning and Peter North both improved the section on pigs and Karl Barton did the same for poultry.

Finally, my wife has read the script, and given me every support.

Jon Newton
Rhos-on-Sea
March 2004

Useful Organic Farming Web Site Addresses

The following short selection of web site addresses will provide the reader with further interesting and useful information as well as some onward links to other web addresses concerned with organic and sustainable farming and food production.

Elm Farm Research Centre
http://www.efrc.com

European Organic Network
http://www.uni-hohenheim.de

IFOAM (International Foundation of Organic Agriculture Movements)
http://www.ifoam.org

Organic Farming Research Foundation
http://ofrf.org

Organic Research
http://www.organic-research.com

Soil Association
http://www.soilassociation.org

The Organic Milk Suppliers Cooperative (OMSCo)
http://www.omsco.co.uk

World Organic Agriculture
http://www.soel.de/english/index.html

USDA National Organic Program
http://www.ams.usda.gov/nop

Yeo Valley Organic
http://www.yeovalley.com

1 Introduction

It is sometimes said that a new cause is best defined by what it is against, rather than what it is for. This has been true of organic farming, that it has been outlined against conventional farming. The purpose of this book is to set out standards for organic farming in its own right and to examine its commercial prospects.

To be successful and to gain adherents, organic farming must be financially viable, and to do this it has to produce quality products at predictable times. This can be achieved only by professional and skilful farmers drawing on scientific wisdom. Supermarkets, where most food is bought, require a constant volume of quality goods. If organic farming is to increase, then this challenge must be met and the consumers must be able to buy organic food at competitive prices. After all, the goals of organic farming – sustainable methods of farming, avoidance of pollution, the welfare of animals and the use of renewable resources – are goals that most humans believe in, at this time of exploitation and overpopulation. It is important that future generations can enjoy the same diversity that is available to us. For this to happen, organic farming must become more widespread and its outputs must be readily available.*

One purpose of this book is to help the organic farmer define how much milk or meat or cereal he can produce from clay or chalk land, and how this will vary according to temperature, rainfall and altitude. Farmers like to do better than their neighbours, or at least as well, producing more bales of hay per ha for example. But, particularly with organic farming, this is likely to be the case only if the

---

* Tamm (2000) 'As a constant supply of the European markets is a prerequisite for the extension of the organic market share, trade may represent a suitable approach to overcome shortcomings of supply – however, efficient and fair trade structures that help farmers to keep prices as well as market supply constant on a multi-region scale are yet to be developed.'

fields are similar in potential. The idea that all fields, irrespective of soil type, rainfall and number of growing days, can be forced to produce the same yield by adding tonnes of artificial fertilisers is alien to the organic farmer. In this book it will be shown that grass yield is related to site class (defined by soil type and number of growing days) and that a farmer with site class 4 land cannot produce as much grass as a farmer with site class 2 land, but that because he can predict the grassland yield, and therefore the amount of milk or meat producible per ha, targets can still be set and met. Furthermore, a knowledge of animal husbandry and organic feeding standards allows animal growth and output to be defined. Thus marketing promises can be made and kept.

It is vitally important for organic farmers that consumers buy organic food and demand that more is available. However, it is difficult to prove that organic produce tastes 'better' than conventional produce. Pastushenko, *et al.* (2000) have demonstrated that intensively managed cattle had higher concentrations of saturated fatty acids than cattle managed according to the standards of organic animal husbandry. In the case of vegetables, it has been shown by Rembialkowska (2000) that organic white cabbage had more vitamin C, more potassium and less calcium than is the case with non-organic cabbage, while organic carrots contained fewer nitrites than conventional carrots. These factors appear relevant to the perceived 'better' taste of organic vegetables.

Consumers are also likely to be influenced in their choice of organic products by the knowledge that an increase in organic farming will improve the welfare of farm animals, and that pollution of the rivers and the countryside generally will diminish.

Although, as King (1926) reminds us, 'the first condition of farming is to maintain fertility', the organic farmer will also need to be aware that an increase in the quantity of organic food bought will come about only if the price is competitive. It will also be part of the farmer's role to increase general awareness as to what 'organic' really means, and to have proof available that the regulations are being kept.

The monitoring of organic farms in Europe and North America is relatively well-established. However, as Herrmann and Heid (2000) pointed out, several European member states are voicing doubts about the reliability of control mechanisms among groups of small farmers in Latin America, Africa and Asia.

According to Sathyanarayana (2000), the isolation of organic farmers in India is immense. This isolation has arisen due to the inability to create a proper network for inspection, certification and procurement.

It is of particular importance that monitoring standards be extended worldwide, since, as Meier (2000) predicts, the organic movement is now on the threshold of changing from a niche to a mass-market.

Organic producers have been seen, and rightly so, as a group of idealists providing food for a minority. For more of the land and countryside to be properly maintained and cared for, this has to change. There is talk of government support through the Organic Aid Scheme and of premiums for organic milk, cereals and meat, and these are very welcome, but for organic farming to become more extensive it has to be able to compete in the market place, which is why consumer support is essential.

Output on a farm is based on a number of systems: for example, milk, egg and barley systems. The regulations published by the Soil Association concerning organic farming prohibit some systems which they regard as cruel or unnatural, and at the same time they encourage organic farmers to diversify as much as possible. This is because the monocultures are regarded as exploitative and likely to favour the build-up of weeds and population explosions of pests. In this book, various systems will be described and their interaction and suitability for particular locations discussed.

It is always expensive for the farmer to change what he produces, because he then has to invest in new machinery or livestock or convert buildings, but new options have to be considered as product demand and supply alters. Unfortunately for the farmer, these changes are rarely predicted and altering an enterprise takes years rather than months.

I visited 15 farms in England, Scotland and Wales five times a year for three years as part of an experiment to measure herbage production on organic farms. During this period many management problems became clear. The farmers involved all agreed to be interviewed further, and the results of these visits and the interviews are related and discussed in Chapter 10. The farms on which the experiment was carried out covered a wide range of farm enterprises and farm size, dealing with dairying, beef production, sheep, wheat, barley, oats, beans, pigs and free range hens. Farm size varied from 24 ha in Wales to 2025 ha in a Scottish glen. Knowledge as to how to succeed, organically, had evolved from mistakes as well as successes. But during the course of these visits it became clear that, technically, organic farming is well within the range of most farmers. It can be done. Furthermore, the livestock look well and grow well, the cereals yield plentiful harvests and the grass–clover swards, particularly the clover, are excellent. There are invariably one or two grumbles at particular regulations, but, given a commitment to the organic cause, these have been overcome.

The problems mainly arise on the sales side. Most farmers, and organic farmers are no different, take great pride and satisfaction in producing healthy stock that look and produce well, and in good hay and cereal crops. That is their area of expertise, and what happens when the product leaves the farm gate should be someone else's domain. It is only the occasional farmer who has the time and the expertise to sell his products directly to the retailer. Cattle can be marketed and cereal sold to the merchant, but when it comes to dealing directly with the supermarket then the farmer is likely to lose out, either because the supermarket dictates what is to be produced on the farmer's farm and how, or alternatively the farmer's profit margin is severely cut back by the buying power of the supermarket. There are ways round this – by selling directly from the farm, or to local butchers and shops – but new health and hygiene regulations have made the first option expensive and time-consuming and the second option may involve too small a proportion of what is produced.

Organic farmers like to sell their products with the organic label, and to know that consumers can buy their food as organic. Far too often they have had to sell their products in the conventional market; quite often this means that they get a higher price. The alternative would be to receive an average price plus a small premium for being organic. Organic farmers want to sell their produce as organic, and to receive slightly more for their product, which could carry the Soil Association symbol for which they have to pay. It must be worth their while, financially, to register with the Soil Association, but on the other hand the consumers should not have to avoid organic food because the price differential is too great. Organic food is unlikely ever to be cheap food, but the cost of supporting the highly worthwhile aims of organic farming must not be too high.

Schmid and Richter (2000) showed that the premium prices for organic food were highest for fruit and vegetables (60% to 70%) and lowest for cheese (20%) and cereal products (31%). The premium prices for organic meat were 52% above conventional and for milk were 42%. The reason given by sales staff for the big price differential for fruit and vegetables was that these products cannot be stored for very long. The authors state that to promote sustained consumer demand the premium price-level should be closer to the conventional price, although consumers are willing to pay some premium. These results are derived from a price survey conducted in 14 conventional retail chains in Germany, Austria, France, Italy, Denmark and the UK.

There has been considerable growth in the organic sector since the first edition of this book was published in 1995. During this period organic sales in Sainsburys in the UK, for instance, have risen from £75 000 per week to £3 million, with an increase from 42 to 630 lines of organic food; but over 75% of this organic food is supplied from abroad, albeit mostly fruit and vegetables. There must, therefore, be considerable scope for more British farmers to become organic and still find a ready market for their product. It is increasingly being recognised that the organic food market in Britain is no longer just a niche market. By 2003 the UK organic market had become the third biggest in the world, worth in excess of £1 billion a year (Soil Association, 2003). The government has increased its financial support for the Organic Farming Scheme, but more money is still required to support farmers during the difficult period of conversion to organic production. The increased interest in becoming organic is shown by the rise in the number of enquiries to the Organic Conversion Information Service (OCIS). In January 1999 this totalled 628 enquiries, more than the total received since the launch of the service in July 1996 and a 25% increase on December 1998 (*Organic Farming*, 1999).

There are now several examples of European countries which have increased their organic agricultural areas, starting from just such a low base as Britain has now. Sweden increased its organic agricultural percentage from 1.6 to 11.2% in four years, Denmark has increased from 1.5 to 5.9% and Austria reached 10.9% by the year 2000.

Hamm and Michelsen (2000) list the proportion of organic land as a percentage of the total agricultural area for eighteen European countries in 1997, with Austria first (10.12%), followed by Switzerland (6.7%), Sweden (6.46%), Finland (4.76%) and Italy (4.08%). At the base of the table were Greece (0.19%), Portugal (0.31%) and the UK (0.34%)*. As far as the organic share of the food market was concerned, four European countries had a value greater than 2% (Switzerland, Austria, Denmark and Sweden), whilst the countries

---

* However, by 2003, the UK had 4% of its farmland under organic management. (Soil Association, 2003)

where least organic food is eaten (less than 0.5%) were Spain, Ireland, Portugal, Greece and the Czech Republic.

The stated reasons for success in increasing the organic share of the food market were:

- strong consumer demand
- high degree of support by food firms
- high proportion of organic food sold through supermarkets
- low price premiums for organic food
- clear, reliable labelling and logos
- promotion of organic food.

By the year 2002 (Yussefi, 2002) there were 17 million ha of land managed organically in the world. The major part of this area is located in Australia (7.7 m ha), Argentina (2.8 m ha) and Italy (more than 1 m ha). Oceania holds 45% of the world's organic land, followed by Europe (25%) and Latin America (22%). In Latin American countries, the organic land area is almost 0.5% of the total area, which represents an extraordinary growth rate from a low level.

For Africa there are only a few figures available but organic agriculture is increasing, mainly to supply organic products to the industrialised countries. In most Asian countries the area under organic management is still very low, but Japan holds the world's third largest market for organic products.

According to estimates, the world retail market for organic food and beverages increased from 10 billion dollars in 1997 to 17.5 billion dollars in 2000. Expected growth rates of organic food are high (10% to 20%), particularly when compared with other food categories.

The extent of this recent growth in organic land area and in organic food consumed is confirmed by many authors for the individual countries. Sylvander and Leusie (2000) talk about 'the craze for organic food that can be found in most industrialized countries'. In Switzerland the number of organic farms quadrupled between 1990 and 1998 (Meyre, 2000). By the year 2000 the Czech republic had 140 000 ha of organic land, which represented 3.4% of the whole agricultural area (Urban, 2000). In the Slovak republic the number of organic farms grew from 37 to 96 between 1991 and 2000, which is an increase of the proportion of organic land from 0.59% to 2.42%.

In India the demand for organic products (spices, medicinal herbs, essential oils, pulses, etc.) has outstripped supply (Sathyanarayana, 2000). In China the estimated value of exported organic products was between 10 and 12 million US dollars in 2000 (Li Zhengfeng & Ding Wei, 2000). For the United States, Jolly (2000) talked about an organic products market worth 3.5 billion dollars, which is growing at a rate of 15% to 20%. This is attracting the attention of serious business interests that will eventually 'mainstream' organic foods.

Kotschi (2000) was concerned that the standards and demand for organic products were dominated by the wealthy northern countries. He recognised that there was a booming international market for organic food and textiles. The green tea from China, coffee from Mexico and cotton from Tanzania are exported to wealthy people in Europe, North America and Japan.

Wier and Smed (2000) point to the great increase in the organic food market in Denmark in the period 1997–8. Sales of organic bread and cereals increased by 143%. During this period relative organic prices decreased, with the exception of meat. Organic dairy prices were the most successful, with a 10% share of the market, followed by bread and cereals (5%), other foods including fruit and vegetables (4%), and the lowest was organic meat with a 1% share.

By 2002 Wier, *et al.* (2002) reported that Denmark had probably the highest per capita consumption of organic products in the world. The reasons for this success were:

- The Danish market for organic food is relatively mature, meaning that it does not suffer from shortages and barriers.
- Organic food is primarily sold through conventional retail stores (mainly supermarkets) and is thus supplied where most of the consumers do their shopping.
- Sales through supermarkets require large and continuing supplies, which in turn tend to reduce average price premiums for organic food.
- Denmark has a very well functioning labelling and certification programme, which the consumers know and trust.

Other countries have ambitious targets for their organic market. Kuhne and Jahn (2002) suggested 20% of the market for organic produce in Germany by the year 2011, whilst Viaux, *et al.* (2002) mention a growth rate of 20% to 30% per year for the organic market in Europe.

Znaor and Kieft (2000) suggested that the approach adopted in Hungary could well increase the organic share in Europe. This approach was to have attractive prices for the export market whilst creating a local demand with slightly lower prices.

As the organic movement grows, so the debate about principles and standards intensifies. There is little doubt that if organic standards were relaxed more producers would become 'organic'. Whether they would make more money is questionable. The higher prices charged for organic products can be justified only if strict principles are adhered to and seen and believed to be upheld. If the large retailers start squeezing organic prices then organic producers may well start to cut corners and, if this were to be exposed by the media, consumer trust would be lost.

On the other hand, there are undoubtedly strict devotees of organic standards who make it clear that they are morally superior and that conventional methods of farming are cheap and nasty. This may be irritating as well as untrue, but there is no doubt that such devotees can be relied on to maintain the highest standards at all times.

In America (Thompson, 1998), these individuals are referred to as 'true naturals', 'young recyclers', 'healthy eaters' and 'affluent healers', while in Ireland (Roddy, *et al.*, 1996) they are 'organic purists'.

Thus, in Britain also, the argument continues between organic farmers with strict principles who quite like being part of a small group doing something difficult and, on the other hand, those who advocate the need for as many organic farmers as possible, so that the British public's requirement for as much organic food as possible, produced in Britain, is met quickly. Two questions then have to be asked. Can a large increase in the production of organic British food, of all kinds, happen only if organic standards are watered down? Or will the prospect of higher prices for organic food, plus the chance of greater profitability and increased financial support from the government, encourage more farmers to become organic, so that the big jump in organic production can occur with annual inspection to ensure that the organic standards are being met by all organic farmers?

Most British organic farmers prefer to sell their produce via local shops. This has three advantages: first, they don't have to compete with the huge buying power of the supermarkets; second, they know that their product will not be moved hundreds of miles, using up valuable energy resources; and, third, they have the satisfaction of feeding people in their own region, which is a reciprocal pleasure, the purchaser probably knowing the farm from which the food comes.

However, supermarkets sell by far the greatest proportion of food, and if organic produce is going to become a significant percentage of the food market then it is essential to sell organic food through supermarkets as well. If organic food is a quality product that has the virtue of being safe and wholesome, then it should be available and affordable to as many people as possible. This has clearly happened in Denmark (see above).

What has changed since 1995 is that large organic businesses, such as Yeo Valley and Chisel Farm, prior to its change of ownership (which are described in detail in Chapter 10), have become increasingly successful at supplying quality products every day of the year, and this is a necessity for supermarkets and their rather pampered customers. There is no conflict between supplying local shops *and* supermarkets with organic food. It should be available at all food outlets.

Do British organic farmers make money? Murphy (1992) in his report following a survey of some organic gross margins believes not, and certainly there are a number of small organic family farms which had financial problems during the recession. In some of these cases it is likely that there were people buying non-viable farms in the first place, which were too small to support even a small family. But, on the other hand, those organic farmers who had viable conventional farms and then became organic are now regularly making more money farming organically than they did when farming conventionally.

Oosting and de Boer (2002) in a 5 year study of dairy farming in the Netherlands showed consistently higher family incomes for organic than for conventional dairy farming. Furthermore, the emission of greenhouse gases and the acidification potential of organic dairy farms were 14% and 40% respectively lower per litre of milk than for conventional dairy farms.

Similar findings on dairy farms in Canada were reported by Stonehouse, *et al.* (2001). The superior economic performance on organic dairy farms was attributed to lower costs of production for almost all material inputs, including dairy herd replacements and livestock feeds. The organic dairy farmers used more land for feed crop production for the dairy cows in order to be as self-sufficient as possible. The conventional dairy farmers imported crop seeds, synthetic chemical fertilisers and pesticides, feedstuffs and herd replacements, with more of their land being devoted to cash crops.

Lah, *et al.* (2002) reported a comparable level of economic efficiency on cattle farms in Italy due to favourable prices and additional subsidies.

Mazzoncini, *et al.* (2000) compared a four year crop rotation (maize–wheat–tomato–durum wheat as main crops and green dwarf beans and spinach as secondary crops) for conventional, low-input, and organic systems. Their conclusions were that low-input and organic systems led to the best economic performances. Similar findings were made in Poland (Kus and Stalenga, 2000) for potatoes and cereals, with the economic system being more efficient for energy use.

Znaor and Kieft (2000) studied organic agriculture in three Danube river countries, Bulgaria, Hungary and Romania, and concluded that the macro-economic benefits of organic agriculture were very pronounced when the environmental costs of nitrogen leaching were taken into account. They feel that this justifies policy incentives for environmentally sustainable agriculture. However, Dabbert (2000) argues that however justifiable direct government support is on environmental grounds, variations in the amount of support between countries tend to distort the competitive position of their organic farmers. He pleads for a consistent policy towards organic farming within the European Union.

Despite these successes, the impression is often given, with certain European countries such as Austria in mind, that Britain's record in the expansion of organic farming is a poor one. When the situation in Britain is compared with the USA, however, a different picture emerges, for Americans are seen, on the whole, as even less interested in where and how their food is produced than are Europeans (Clancy, 1997).

The separation between food consumers and producers in America has been ascribed to several factors. With the development of an urban industrial society, people became dependent on unknown producers for much of their food supply. As transport systems improved (railways, roads and canals) it became easier to produce food in regions whose climates allowed year-round production.

Supermarkets developed in the 1920s, gradually replacing small markets. More people owned cars, and out-of-town supermarkets were built with greater space for merchandising and parking. There was also an increase in the globalisation of food resources. Because of this, many people in the United States lost the concept that nature dictates the food available in a particular area. More recently, the chemical revolution in agriculture has damaged consumers' faith in the safety and value of food, and there has also been the development of synthetic dairy and meat products, with no obvious, or actual involvement of a farmer at all.

The modern north American dogma of maximising production and the need to 'get big or get out', coupled with the fact that the largest profits in the food system are made by the processors, has made the outlook for the farmer bleak in many respects.

The ever-increasing power of the food industry makes a mockery of free enterprise, as farmers find markets difficult to enter (Merrigan, 1997). One supermarket in America advertises itself as 'supermarket to the world'. Farms have got bigger in America, particularly in the livestock industry, with pigs housed in lots of 3000 or more, cattle herds numbering tens of thousands and broiler units of 500 000 hens: these are in danger of becoming the norm.

Small farmers in the USA find it difficult to use the services of packing houses and distributors, either because they are considered too small to bother with or because such services are owned by their large competitors. The four largest firms in the meat-packing industry control more than 80% of the beef market. There are other concerns over the power of the large corporations. Industry giants get free access to shelf space in supermarkets for popular manu-factured items, whereas smaller companies must pay fees for similar access.

But this trend can and must be changed. Agriculture is one of the many industries and activities that have degraded the environment, and modern science has alerted us to the fact that 'ecology is destiny' (Freedgood, 1997).

Consumers will have to 'choose to have less choice'. It has to be recognised that much of the choice in the food market is spurious, an illusion of artificial colours, flavours and additives. It is also wasteful of the fossil fuels used to transport foods all over the world and harmful to local farmers and communities. In the USA (Stauber, 1997) rural America is no longer agricultural America, with only about a quarter of rural counties, mostly in the Midwest, dependent on agriculture.

There are pleasures in eating local food and knowing that the farmers have considered how the natural world functions, have minimised soil erosion and water degradation and have cared about their livestock. In America today, the problems of soil erosion and water degradation are of particular relevance.

Two main suggestions are made for reducing the destructive land-use patterns in parts of America (Merrigan, 1997). First, the government must intervene to change the production regions for several crops and, second, crop rotation must be compulsory so soil erosion and the use of agrichemicals can be reduced.

An example of this is the discouragement of fruit and vegetable production in Florida, because although the humid climate is suitable for the crops it is also a perfect climate for many pests that have to be controlled by heavy application of fungicides and other toxic chemicals. Furthermore, without subsidies there would be far less irrigation: irrigation now consumes precious water to produce such low value crops as hay, maize and wheat, which could otherwise be grown in regions that do not depend on irrigation.

Monocropping, which is known to require heavy applications of pesticides and fertilisers, must give way to crop rotations that increase soil organic matter, replace nitrogen and break pest cycles. Six suggestions are made in favour of rotations: farmers have to be persuaded to plant less profitable rotational crops by encouraging eating patterns based on a more varied diet; there should be subsidies for unprofitable crops that are beneficial in a rotation; farmers who rotate crops would receive farm credits; legume planting should be encouraged at the expense of fertiliser application; research programmes that are funded for only three years should be extended so that longer rotations are studied; finally, all areas that are at risk environmentally should be encouraged to use crop rotations.

Diversity should be encouraged in agricultural methods, in crop and livestock varieties, in land and in the farmers themselves. New entrants are needed in the farming profession. Some of the best research pioneers are organic farmers. Yet the system in America does not encourage organic farming. Many organic farmers find it difficult to obtain farm credit because bankers require chemical production methods, to ensure high yields. Organic farmers complain that contract growing for large enterprises requires them to conform to a schedule set by firms which buy their produce. Even the research system is biased against organic production, with less than 1% of the United States Department of Agriculture (USDA) budget being spent on organic farming.

Merrigan (1997) has come to the conclusion that:

'The Justice Department, Federal Trade Commission, and USDA should be directed to ensure that agriculture has a multitude of competitive markets. I concur with those who argue that rather than promoting a global supermarket run by multinational corporations, the role of government should be to ensure creation of a globe of agricultural villages.'

References

Clancy, K. (1997) Reconnecting Farmers and Citizens in the Food System. In: *Visions of American Agriculture* (ed. William Lockeretz), pp. 47–58. Iowa State University Press, Ames, USA.

Dabbert, S. (2000) Organic farming and the Common Agricultural Policy: A European Perspective. *Proceedings of the 13th International IFOAM Scientific Conference*, Basel, Switzerland, pp 611–14.

Freedgood, J. (1997) Farming to Improve Environmental Quality. In: *Visions of American Agriculture* (ed. William Lockeretz), pp. 71–90. Iowa State University Press, Ames, USA.

Hamm, U. & Michelsen, J. (2000) Analysis of the organic food market in Europe. *Proceedings of the 13th International IFOAM Scientific Conference*, Basel, Switzerland, pp 507–510.

Herrmann, G. & Heid, P. (2000) The weakest go to the wall: Inspection and certification overkill for small farmers. *Proceedings of the 13th International IFOAM Scientific Conference*, Basel, Switzerland, pp 559–561.

Jolly, D. (2000) From cottage industry to conglomerates: The transformation of the US organic foods industry. *Proceedings of the 13th International IFOAM Scientific Conference*, Basel, Switzerland, pp 511–12.

King, F.H. (1926) *Farmers of Forty Centuries.* Jonathan Cape, London.

Kotschi, J. (2000) Poverty alleviation in the South. Can organic farming help? *Proceedings of the 13th International IFOAM Scientific Conference*, Basel, Switzerland, pp 652–5.

Kuhne, S. & Jahn, M. (2002) Regulations on the use of plant protection and plant strengthening products in organic farming in Germany. *Proceedings of the 14th IFOAM Organic World Congress*, Victoria BC, Canada, p. 45.

Kus, J. & Stalenga, J. (2000) Comparison of economic and energy efficiency in ecological and conventional crop production systems. *Proceedings of the 13th International IFOAM Scientific Conference*, Basel, Switzerland, pp 395.

Lah, B., Bovolenta, S., D'Agaro, E., de Ros, G. & Menta, G. (2002) Organic cattle farming in Italy: Four case studies in the Italian Simmental breed. *Proceedings of the 14th IFOAM Organic World Congress*, Victoria BC, Canada, p. 83.

Li Zhengfeng & Ding Wei (2000) Organic certification for small farmers in China. *Proceedings of the 13th International IFOAM Scientific Conference*, Basel, Switzerland, pp 572.

Mazzoncini, M., Bonari, E., Silvestri, N., Coli, A., Belloni, P. & Barberi, P. (2000) Agronomic and economic evaluation of conventional, low input and organic farming systems in central Italy. *Proceedings of the 13th International IFOAM Scientific Conference*, Basel, Switzerland, pp 393.

Meier, T. (2000) Organic production and fair trade products: bringing together social and ecological issues. *Proceedings of the 13th International IFOAM Scientific Conference*, Basel, Switzerland, pp 563.

Merrigan, K.A. (1997) Government Pathways to True Food Security. In: *Visions of American Agriculture* (ed. William Lockeretz), pp. 155–74. Iowa State University Press, Ames, USA.

Meyre, S. & Steinhofel, H. (2000) How are organic foods distributed in Switzerland? *Proceedings of the 13th International IFOAM Scientific Conference*, Basel, Switzerland, pp 695.

Murphy, M.C. (1992) *Organic Farming as a Business in Great Britain.* Agricultural Economics Unit, Department of Land Economy, University of Cambridge, Cambridge.

Oosting, S.J. & de Boer, I. J. M. (2002) Ecological, Socio-cultural and Economic Sustainability of Organic Dairy Farming in the Netherlands. *Proceedings of the 14th IFOAM Organic World Congress*, Victoria BC, Canada, p. 112.

Organic Farming (1999) Conversion enquiries reach new heights. *Organic Farming*, **61**, 3.

Pastushenko, V., Matthes, H-D., Hein, T. & Holzer, Z. (2000) Impact of cattle grazing on meat fatty acid composition in relation to human nutrition. *Proceedings of the 13th International IFOAM Scientific Conference*, Basel, Switzerland, pp 293–6.

Rembialkowska, E. (2000) The nutritive and sensory quality of carrots and white cabbage from organic and conventional farms. *Proceedings of the 13th International IFOAM Scientific Conference*, Basel, Switzerland, pp 297.

Roddy, G., Cowan, C.A. & Hutchinson, G. (1996) Consumer attitudes and behaviour to organic food in Ireland. *Journal of International Consumer Marketing*, **9** (2), 41–63.

Sathyanarayana, M.G. (2000) Organic marketing in India – prospects and problems. *Proceedings of the 13th International IFOAM Scientific Conference*, Basel, Switzerland, p 533.

Schmid, O. & Richter, T. (2000) Marketing measures for selling organic food in Europe retail chains – key factors in success. *Proceedings of the 13th International IFOAM Scientific Conference*, Basel, Switzerland, pp 519–522.

Soil Association (2003) *Organic food and farming report.* Soil Association, Bristol.

Stauber, K. (1997) Envisioning a Thriving Rural America through Agriculture. In: *Visions of American Agriculture* (ed. William Lockeretz), pp. 105–18. Iowa State University Press, Ames, USA.

Stonehouse, P., Clark, E.A. & Ogini, Y.A. (2001) Organic and conventional dairy farm comparisons in Ontario, Canada. *Biological Agriculture and Horticulture*, **19**, 115–125.

Sylvander, B. & Leusie, M. (2000) Consumer trends in organic farming in France and Europe: vulnerability of demand and consumer loyalty. Towards a learning-based marketing. *Proceedings of the 13th International IFOAM Scientific Conference*, Basel, Switzerland, pp 546–9.

Tamm, L. (2000) The future challenges and prospects in organic crop protection. *Proceedings of the 13th International IFOAM Scientific Conference*, Basel, Switzerland, pp 106–9.

Thompson, G.D. (1998) Consumer demand for organic foods: what we know and what we need to know. *American Journal of Agricultural Economics*, **80** (5), 1113–18.

Urban, J. (2000) 10 years of organic farming in the Czech Republic. *Proceedings of the 13th International IFOAM Scientific Conference*, Basel, Switzerland, p 693.

Viaux, P., David, C., Bellocq, G. & Sancho, E. (2002) The development of Organic Cereals in Europe: A comparative analysis in France, Denmark and Italy. *Proceedings of the 14th IFOAM Organic World Congress*, Victoria BC, Canada, p. 126.

Wier, M., Hansen, L.G., Andersen, L.M., Milloock, K. & Browning, M. (2002) Consumer preferences for Organic Foods. *Proceedings of the 14th IFOAM Organic World Congress*, Victoria BC, Canada, p. 189.

Wier, M. & Smed, S. (2000) Modelling Consumption of Organic Foods. *Proceedings of the 13th International IFOAM Scientific Conference*, Basel, Switzerland, p 550.

Yussefi, M. (2002) Organic Agriculture develops rapidly world-wide. *Proceedings of the 14th IFOAM Organic World Congress*, Victoria BC, Canada, p. 195.

Znaor, D. & Kieft, H. (2000) Environmental impact and macro-economic feasibility of organic agriculture in the Danube River Basin. *Proceedings of the 13th International IFOAM Scientific Conference*, Basel, Switzerland, pp 160–3.

2 Grassland Productivity

The basis of successful milk, beef and sheep production is grass, particularly for the organic ruminant, for whom roughage must form 60% or more of its diet. The balance between grazing and conservation as silage or hay varies with the enterprise. In dairying there is a greater emphasis on conservation because cows are likely to be fed silage for six months of the year, whereas sheep may well be grazing outside throughout the winter, and offered hay or silage only when there is snow on the ground.

In 1996 the total amount of organic grassland in the 15 European Union countries was 877 000 ha (Foster and Lampkin, 1999), with the largest areas in Austria (237 000 ha) and Italy (201 000 ha). Great Britain with 39 000 ha ranked seventh. As a percentage of total certified organic and in-conversion land area, the 18 countries (EU + 3) had 56.2% organic grassland, as opposed to organic arable. The countries with the highest proportion of organic grassland (again, as opposed to organic arable) were Switzerland (91.25%), Norway (89%) and then Great Britain (78.88%).

Grass production is not a constant. It varies between fields, depending on soil type, soil nutrients, rainfall, latitude, altitude, slope, aspect, herbage constituents and management. Furthermore, the rate of growth varies between months and years, growth being favoured by length of day, temperature and moisture. The less contrast there is in rate of growth between months, the easier management becomes. In favoured parts of New Zealand grass will grow in every month of the year, making decisions as to stocking rate much easier. Where the grass growth curve is compressed, as in northerly latitudes, then the decision as to how many animals to carry is more difficult, particularly if there is variation in production between years. Most farmers prefer to under-stock and conserve excess herbage as silage or hay, rather than risk being caught with no grass for their animals for lengthy periods.

Grassland in Britain

There are approximately 18.5 million ha of agricultural land in Britain (Fig. 2.1), of which 11 million ha is permanent pasture (60%) and 1.7 million ha is temporary grassland (9%). The distinction between permanent pasture and temporary grassland is the time since it was sown. The dividing line is normally five years: thus grassland that was sown six or more years ago is normally classified as permanent pasture. Permanent pasture can then be subdivided into 5.2 million ha of permanent grassland and 5.8 million ha of rough grazing. The distinction between permanent grassland and rough grazing is not absolute, but the assumption is that rough grazing will be within land-use capability classes 5–7, ranging from a severe to extremely severe degree of limitation for agricultural use. Factors preventing improvement are high altitude, extreme exposure, severe climate, excessively steep slopes, poorly drained boggy land, very stony parent material or bare rock.

Reasons for predicting the productivity of organic grassland

Before considering methods available for predicting yield from grassland, it is useful to consider why an accurate prediction is important. The main reason is that there is at least a fourfold differ-

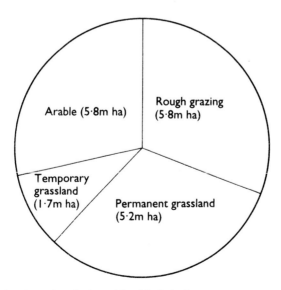

Fig. 2.1 The division of agricultural land in Britain.

ence in yield between the best and the worst fields, and this variation is not significantly affected by management. This large difference has crucial implications for the number of animals that can be grazed on a field and therefore per farm. It is clearly absurd to expect the same herbage production from a steep hillside as from a lush dairy pasture.

The second important point is that although factors that affect herbage production, such as climate, latitude and aspect, are outside the farmer's control, other factors, such as soil nutrient level and grazing management, can be controlled by the skilful farmer.

The third important factor is that the method of prediction should not be so expensive that the farmer is unlikely to want to pay for the answer, as might be the case if the prediction had to depend on a whole series of detailed laboratory analyses.

Finally, there is the question of the accuracy of the prediction. It is highly likely that any increase in accuracy will also increase the cost. For a start, the smallest animal likely to be used in grazing is a sheep, which will consume 0.5 t of dry matter per annum. This means that there is little point in trying to predict within less than 0.5 t of dry matter. With an all-dairy farm, accuracy can be less precise, because a dairy cow is likely to eat 4.2 t of dry matter per head per annum.

Factors influencing herbage yield

Soil type and depth
Soils consist of differing proportions of clay, silt or sand. Clay, which has the smallest particles, can be difficult to manage, and suffers from drainage problems, but on the positive side it is likely to be more resistant to drought and to hold on to nutrients, which are fundamental for plant growth. Sand, on the other hand, which consists of large particles, is freely drained and nutrients are more easily washed down, out of reach of plant roots. Silt or loam is intermediate between the two. A further type of soil is one with a high proportion of chalk: this is characterised by a high pH and free drainage, but it is more likely to be nutrient-rich than sand.

Deep soils are more fertile than shallow soils (shallow is defined here as less than 30 cm).

Latitude
In Britain the further north the farm is, the colder the annual temperature and the shorter the growing season for grass and clover.

Altitude

Crompton (1958) made the point that on the western seaboard of Britain it is only in exceptional years that even the January temperatures fall below 6°C, and some grass growth is therefore possible throughout the year, whereas at the summit of Ben Nevis (1342 m above sea level) even the July mean temperature does not normally rise to 6°C and there is no growing season at all for ordinary plants.

Hunter and Grant (1971) demonstrated that with some grass cultivars yield declined by approximately 2% for every 30.5 m increase in altitude, and that the development of flowering was delayed by 1.3 days per 30.5 m increase in altitude.

It is generally accepted that below 6°C grass is unlikely to grow, and this is known as the cardinal temperature. Alberda (1965) stated that at temperatures above 11°C grass growth is independent of increase in temperature and that under optimal nutritional conditions growth is then determined by light intensity.

The temperature range in Britain can be classified into three distinct ranges:

(1) Below the cardinal temperature.
(2) The temperature responsive range – temperatures above 6°C but below 11°C. (Within this range temperature variations will have their maximum effect on growth.)
(3) Plateau range – temperatures above 11°C. Where the mean temperature exceeds 11°C, growth is accelerated and the limiting factors are light, moisture or nutrients.

Hunter and Grant (1971) showed that the magnitude of the altitude effect on yield varied with season. In spring, yields were decreased by some 5% for every 30.5 m rise in altitude and in autumn by 1.8%. In summer, yield trends were non-significant or reversed, highest yields sometimes occurring at the higher altitudes. This result was related to the development of moisture stress at lower altitudes.

A similar point was made by Crompton (1958). The remarkable change with altitude is accentuated by the great variation in annual rainfall, which is commonly below 760 mm in the lowlands but rises to as much as 2520 mm and even 5000 mm in the hills. Moreover, there is a still greater contrast between the 1500 mm typical of a large part of the hill country and the 760 mm of the lowlands: evaporation and transpiration remove almost 500 mm, leaving an excess of only some 260 mm in the lowlands, whereas the surplus with a 1500 mm rainfall is at least 1000 mm; thus, with twice the annual rainfall there is the beneficial effect of four times the amount of excess water.

Rainfall

The main grass growing season in Britain is from April to September, and it is rainfall during this period that is of most importance, although sufficient winter rainfall will ensure that soil water is plentiful at the start of the growing season.

Water is absorbed by the plant through the roots and transpired through the leaves, mainly during daylight. The transpiration stream provides soil nutrients in solution and keeps the grass plants turgid. This turgidity ensures optimal photosynthesis by the leaves.

As far as grass growth is concerned, the most important source of water is that available to the roots, and this is known as the available water capacity (AWC). The remainder is unavailable and while this is highest in heavy clays and clay loams, these soils also have the highest AWC. The AWC of soils is improved by an increase in their organic matter content, which improves the soil's capacity to absorb and hold water and also benefits soil structure.

The relationship between rainfall and AWC is complicated by evapotranspiration, which is the rate of evaporation from sward foliage before the water can reach the soil, which will be highest in hot weather, and also by water run-off, which is high on steep land and on sloping, compacted land. If the soil is poorly drained as is likely to be the case with heavy clay land, or soils with impeded drainage, caused by an impermeable subsoil, then it becomes waterlogged. This leads to poor aeration and lack of oxygen, and the nitrogen present in the soil becomes unavailable. This in turn leads to inhibited root growth, loss in grass production and encroachment by rushes. Furthermore, in wet, waterlogged sites there is the risk of denitrification (the loss of inorganic nitrogen to the atmosphere as ammonia), and the breakdown and mineralisation of organic residues are slowed down so that nutrient recycling becomes inefficient.

Long-term grasslands provide plenty of organic matter for humus formation. If the soil is not acidic or waterlogged, the soil fauna and flora create humus and the subsequent useful development of a porous crumb and granular structure. Porosity encourages root growth and the uptake of soil nutrients.

Grass roots can grow and absorb water down to 1 m, or deeper if soil structure is good, although subsoil water is usually poorer in nutrients than topsoil water (Frame, 1992). Transpiration is highest when grass growth is most vigorous, so water deficit on light soils soon limits grass growth in areas of low rainfall. As the roots take up the available water from the soil pores, the soil gradually dries out from the top downwards. Unless this water is replaced by rainfall or upward capillary movement, the soil develops a water deficit. When this deficit reaches a critical stage, transpiration is reduced and grass growth virtually ceases.

The factors considered above are all factors over which the farmer has no control, but they are important in determining herbage production. The next factors to be considered can be altered by management.

pH

Soil pH is easily tested for and determines the availability of nutrients and the success of white clover. Very acid soils (below pH 5.0) will cause a deficiency of the trace elements iron, boron, copper and molybdenum and conversely will cause injury to plant growth by increasing the availability of aluminium and manganese to toxic levels. Over-liming, on the other hand, which can raise the pH above 6.5, will reduce the availability of certain essential elements such as phosphorus, manganese and boron.

Forage legumes, of which white clover is the most important, are particularly sensitive to calcium deficiency and will not thrive. Ideally, soil pH should be maintained between 5.8 and 6.5. Heavy-textured soils such as clay require more lime than sandy soils to raise pH, because the higher levels of clay and organic matter act as a buffer against change, and the same is true of peaty soils.

Basic slag used to be a popular fertiliser for supplying phosphate, but it also had liming value and contained some trace elements. However, the older steel-making processes are now out-of-date and basic slag is much less freely available. The consequence of this is that soils in Britain are becoming more acidic.

Soil nutrients

Regular soil analysis is also important for checking the levels of the important nutrients – phosphorus, potassium, calcium and magnesium. Care should be taken to confirm the method of analysis, because the Scottish system and that carried out by organic laboratories and by Agricultural Development and Advisory Service (ADAS) laboratories have different standards. For instance, on the Scottish system, 'very low' for potassium (the lowest level) is less than 40, whereas the

ADAS lowest level of '0' for potassium is less than 60. As the level of soil nutrients decreases, so herbage production is reduced. What is not so clear at the moment is whether very high levels of phosphorus and potassium increase herbage production.

Legume content

It is universally agreed that the presence of legumes in the sward increases herbage production, mainly because of their power to fix atmospheric nitrogen and to enrich the soil, and thus other plants (e.g. grasses), with this nitrogen when released by the roots and root nodules. The most important legume in this context is white clover because it is perennial. Red clover is also of great value in leys, but does not persist for more than three or four years. Lucerne and sainfoin tend to be grown as specialist crops.

The contribution of white clover to yield is greatest in the absence of added artificial nitrogen fertilisers, which are banned anyway on organic farms. It is generally accepted that white clover will double or treble the yield of grass alone, as well as increasing digestibility and intake, but there is no clear relationship between proportion of clover in the sward and increase in yield, mainly because of the unpredictability of nitrogen release into the soil by the clover. Also, a pure white clover sward will yield less dry matter per ha per annum than a grass/ white clover sward, but the turnover point is not well defined.

King (1926) discussing the highly intensive systems of farming in China, Japan and Korea repeatedly mentioned the importance of clover as a green manure before rice. The clover was cut, saturated in mud, allowed to ferment for 20 to 30 days and then put on the field.

In a comparison of the yield and nitrogen fixing ability of white clover, red clover and lucerne, Loges, *et al.* (2000) found higher yields of harvestable dry matter and therefore of nitrogen fixation for red clover and lucerne than for white clover, but white clover, of course, is a perennial.

Leys and permanent pasture

It is generally assumed that leys will out-yield permanent pasture. The main reason for this assumption is that by definition leys are grown on fields that can be ploughed, whereas permanent pasture is mainly on steep, stony or inaccessible land which is difficult or impossible to plough. It is not therefore the sward constituents that are determining the difference in yield, but the soil type, depth, pH and nutrient status.

Where temporary and permanent grassland have been compared on similar sites, there is likely to be no difference in dry matter yield.

Grassland management

More will be written about this in later chapters but for the purpose of predicting herbage production, the critical point is leaf area index (LAI). Optimal photosynthesis by the grass plant depends on its being able to intercept maximum light. This has led to the concept of LAI, which is a description of how well an area is covered with leaves. LAI 1 is low and LAI 5 is high. If management of stock leads to overgrazing, with consequent short swards (below LAI 3 and height 3 cm), then photosynthesis and growth will be reduced. In a similar way, after a field has been cut for silage, photosynthesis by the stubble is low until such time as the leaf canopy has been restored.

Under-grazing on the other hand will result in decreased utilisation of herbage by the animal, due to large amounts of senescent and wasted material. When a sward reaches full canopy, the rate of senescence will equal the rate of new leaf appearance and there will be no further increase in the weight of the standing crop (Fig. 2.2).

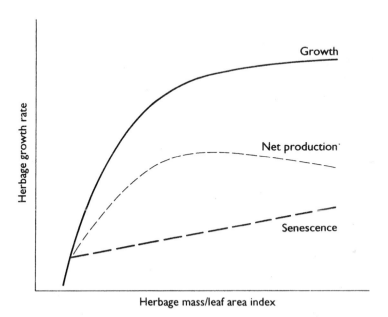

Fig. 2.2 The influence of herbage mass on net production.

Methods of predicting grassland production

The value of predicting grassland production accurately is that the optimal stocking rate for livestock, whether of dairy cows, beef cattle, sheep, pigs or poultry, can then be achieved. The optimum can be defined as the amount of stock that can be kept in an average year without the animals going short of food and without grass being wasted by becoming senescent.

The production of cereal and root crops is directly measured by weighing the harvested yield. The yield of grassland that is cut for hay or silage can also be measured, if the dry matter is known; but there is still no measurement for the period when that field is grazed. During grazing, animals can be weighed regularly and an estimate of the herbage required to satisfy weight change and maintenance can be calculated approximately. But it is an approximation and this, added to an estimate of the utilisation efficiency of the grassland by the animal, makes an accurate figure for herbage production under grazing difficult.

Evans (1951) described land potential and measured it in terms of sustained yield of oats. The determining factors influencing yield were, as we have seen, soil type and climate, particularly rainfall. He then went on to calculate whether a particular farmer was achieving the optimal output, or whether he was under-achieving, or exploiting his land by over-production; in the one case he was making less money than he could have and in the other case productivity would have quickly fallen as the soil lost both structure and nutrients.

Harrod (1979), nearly 30 years later, produced a table (Table 2.1) to define grassland yield categories, which he subsequently refined to include trafficability and the risk of poaching. Unfortunately the

Table 2.1 Grassland yield categories.

Dryness sub-class		Yield categories			
		Dry lowlands	Moist areas	Short growing season	
Increasing dryness	a	a	a	b	Increasing yield
	b	b	a	c	
	c	c	b	d	
	d	d	c	d	

(Harrod, 1979)

categories were never defined in terms of grass yield of dry matter per ha. This was rectified in the booklet *Milk From Grass* (Thomas & Young, 1981), in which grass yields were assigned to particular site classes (Table 2.2) based on the amount of nitrogen fertiliser added. The potential yield of grass with 0 nitrogen added was as shown in Table 2.3. These values were later changed in the second edition (Thomas & Young, 1992), as shown in the same table.

Table 2.2 Site classes.

Soil texture	Average April–September rainfall		
	> 400 mm	300–400 mm	< 300 mm
Clay loams and heavy soils	1	2	3
Loams, medium-textured soils and deeper soils over chalk	2	3	4
Shallow soils over chalk or rock, gravelly and coarse sandy soils	3	4	5

Add 1 for northern areas, i.e. Scotland, or if over 300 m elevation.

Table 2.3 Potential grass production with no added nitrogen (t dry matter per ha).

Site class	1981 edition	1992 edition
1	3.2	5.6
2	2.8	5.6
3	2.3	5.6
4	2.0	5.6
5	1.7	5.1

Site class as defined in Table 2.2.
(Thomas & Young, 1981, 1992)

In the 1981 edition, site class has affected grass production, but in the 1992 edition yield is assumed to be the same for site classes 1–4 and to be much higher than in the 1981 edition. The plots that these figures were based on were arable fields, some down to grass and then cut with a mower to estimate yield. Although no fertiliser was added in the year of measurement, they cannot be termed 'organic'.

A method of predicting organic grassland productivity

An experiment was carried out in 1992 and 1993 on a range of organic farms to determine whether reliable figures of herbage yield can be produced based on site class classification. The site class scheme is shown in Table 2.4. Soil type is based on simple analysis carried out in the laboratory, although experts can assess soil texture by feel. The number of grass growing days (Fig. 2.3) is based on a map, reproduced from Lazenby and Doyle (1981), in which soil temperature has been adjusted for drought and altitude. Dry matter production was measured on two fields on 16 farms, making 32 sites, in 1992 and in 1993. The organic farms in England, Scotland and Wales covered a wide range of soil types, climatic conditions and farming enterprises. Results for 1992 and 1993 are shown in Table 2.5.

Table 2.4 Determination of site class.

	Number of grass growing days		
Soil type	> 280	220–280	< 220
Clay loams and heavy loams	1	2	3
Loams, medium-textured soils and deeper soils over chalk	2	3	4
Shallow* soils over chalk or rock, gravelly and coarse sandy soils	3	4	5

* Shallow is 30 cm or less.
Add 1 if over 300 m.
(Newton, 1992)

Table 2.5 Dry matter production on organic farms by site class.

Site class	Dry matter yield (t DM/ha/yr)	Dry matter yield (t DM/ha/yr)
	1992	1993
2	14.1 (6)	15.7 (8)
3	11.9 (10)	14.6 (12)
4	10.2 (11)	12.8 (8)
5	6.1 (1)	12.3 (4)

Number of sites in brackets.

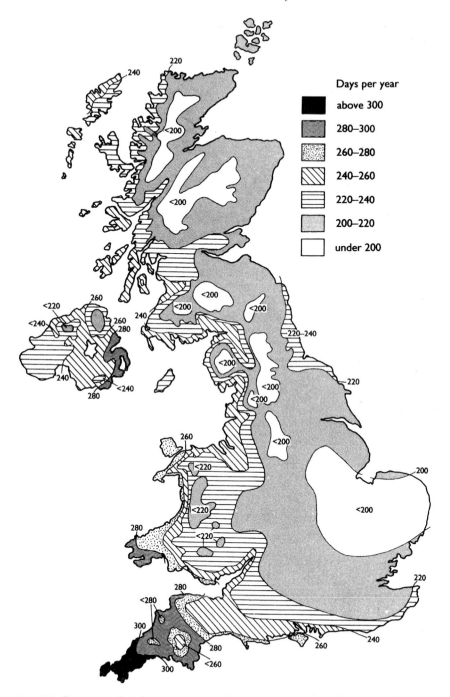

Fig. 2.3 Grass growing days per year (soil temperature adjusted for drought factor and altitude) (Lazenby & Doyle, 1981).

From Table 2.5 it is clear that the assigning of site class numbers is a reasonable basis for prediction. It is apparent from talking to farmers that there is a certain degree of satisfaction if the measured yield in their fields is above average for that site class, but that they will take action if their fields show up as below average. The most usual cause of low production is a shortage of phosphate (P) or potassium (K), no clover, a particularly low pH, or the siting of the field on a pronounced slope. The clear value of the results is that they provide a bench-mark against which to judge other organic fields.

When considering stocking rate, and here all the grassland on the farm needs to be assessed, it must be remembered that annual yield will vary between years (see Table 2.5) and, more importantly, that the seasonal pattern of yield will vary between years, mainly being influenced by rainfall. The response of most farmers is to under-stock slightly, thus building up a reserve of either silage or hay. This can then be offered to livestock in times of shortage. Organic farmers have to feed organic hay or silage and this is not always readily available at an economic price.

Halberg and Kristensen (1997) built a model to predict the yields of various crops, based on data from Danish organic and conventional mixed dairy farms. They found that the yield of grass/clover on organic farms was between 12% and 17% lower, depending on soil type and irrigation, but stated that this was not surprising considering that the grass/clover on the conventional farms received more than 200 kgN per ha per annum as chemical fertiliser. They warned against using their model to predict actual yields on other farms because of the large variation in yield between the individual farms in their survey. In Switzerland, Men, *et al.* (2000) found only a 5% yield reduction for organic grass/clover fields.

The effect of white clover

The presence of white clover is fundamental to organic farming. It increases the dry matter yield of herbage swards by fixing atmospheric nitrogen, which is then released into the soil and becomes available to the roots of the surrounding grasses. White clover is also an excellent feed for livestock, since it has a higher protein content than grass and maintains its digestibility at a higher level with maturity than grass. Because it has fewer cell wall constituents than grass, it has higher intake characteristics, which lead to more milk and more growth.

The problem in predicting the yield characteristics of a grass/white clover sward is that the relationship between yield and proportion of white clover in the sward is not linear. Grass/white clover swards yield more dry matter than either grass or white clover swards on their own, but the point where increasing the amount of white clover in the sward actually decreases total yield has not been defined experimentally.

In Denmark, Hogh-Jensen (2000) studied the effect of white clover on the nutritional content of the companion perennial ryegrass and found that white clover increased the K and P content in the ryegrass.

Curll, *et al.* (1985) measured an increase in total yield from 6 t dry matter per ha with 10% white clover in the dry matter up to 11 t dry matter per ha with 45% white clover (Fig. 2.4). The increase in total yield was mainly caused by the increase in clover yield; the grass yield remained more or less static, with clover levels above 20%. Reports from other experiments show considerable variation in the response. For this reason it can be stated only that white clover increases total herbage yield, but not at very high levels of clover, and that white clover always improves milk yield and liveweight gain compared with grass alone.

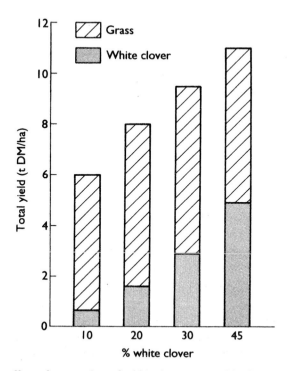

Fig. 2.4 The effect of proportion of white clover on total herbage yield.

Herbage utilisation

Utilisation is a measure of the efficiency with which the herbage that is grown is consumed by the animal. With under-utilisation, clumps of mature herbage develop which, as they senesce, become of increasingly low digestibility, and are then rejected by the animal. When a sward is over-utilised, then sward height is kept at 3 cm or less, leaf area index is continually sub-optimal, less herbage is therefore grown, the sward becomes open and root development is reduced and shallow.

Peel, *et al.* (1988) measured herbage production, milk yield, liveweight change and amount of silage made on six dairy farms in the south of England. The apparent efficiency of utilisation averaged 67%, with a range of 51% to 83%, the low values being on the badly drained farms where poaching was a likely problem, particularly with heavy animals such as dairy cows. The efficiency of utilisation is clearly a most important factor that influences economic productivity. With 100 ha of grassland and a potential herbage production of 12 t dry matter per ha, but with an efficiency of utilisation of 50%, then only 600 t is consumed by the animal; the consequent size of the potential dairy herd will be only 142 cows. If the efficiency of utilisation is increased to 70%, which can be done at no extra cost, then the dry matter available will be 840 t and the dairy cow number can be increased to 200 cows, an increase of 58 animals, brought about simply by better management.

Grazing cattle pastures with sheep will increase the output: first, because sheep will graze closer to dung pats and so waste less grass, and, second, because they can graze the pastures for longer in the autumn and winter without damaging the sward. In fact, the sward will benefit from sheep grazing, the grass will be encouraged to tiller and make a denser sward, the grass available in the spring will be younger and therefore more digestible and, if the sward contains white clover, it will benefit from being opened up by tighter grazing. The effect of a short sward in the autumn (3–5 cm) is to allow more light to reach the white clover stolons; this encourages axillary buds, which will increase the spread of white clover in the spring.

Silage and hay production

Silage- and hay-making are very important components of success on organic farms, not just because good quality silage and hay are critical if minimal concentrates are to be fed without jeopardising animal health and welfare, but also because it is neither easy, nor economic, to buy in organic hay or silage.

The techniques of making quality silage without additives and hay have been described in other books (Newton, 1993), but the likely yields from organic farms, by site class, have not been presented before. The figures shown in Table 2.6 have been calculated from standing crop measurements on organic farms. It has been assumed that the field has not been grazed since February, and that the losses in silage- and hay-making are 20% compared with the weight of the standing crop. The reduction occurs through respiration and mechanical loss. It does not include losses in storage. The standard error of 10% represents annual variation.

Table 2.6 Silage and hay yield according to site class.

Site class	Silage cut 1 June (t DM/ha)	Haylage cut mid-July (big bales/ha)	Hay cut mid-July (t fresh weight/ha)
2	3.4 ± 0.4	36 ± 4	6.7 ± 0.7
3	3.1 ± 0.4	35 ± 3	6.5 ± 0.7
4	2.2 ± 0.3	28 ± 3	5.1 ± 0.5
5	1.8 ± 0.2	23 ± 2	4.2 ± 0.4

The date for cutting silage, haylage or hay will vary between lowland and upland farms and between years. The dairy farmer is more likely to cut early for higher quality silage than the beef and sheep farmer, who is more likely to want more dry matter with less emphasis on quality.

To calculate the fresh weight of silage made, divide by the dry matter percentage. Thus, a site class 2 field yielding 3.4 t of silage dry matter per ha of 20% dry matter, ensiled at the beginning of June will weigh 17 t fresh weight. The number of bales of haylage made in mid-July assumes a dry matter weight per bale of 156 kg (Hutchinson, 1982). However, the wet weight, % dry matter and dry matter weight per bale can all vary. The range of dry matter weights per bale can vary between 100 and 208 kg. The heavier the dry matter weight, the fewer the number of bales per ha and vice versa. Wet weight can vary between 330 and 500 kg, and % dry matter from 25 to 63%. The weight of hay conserved in mid-July is fresh weight, with an assumed

85% dry matter. To translate the fresh weight of hay made into small bales, divide by 0.02. Thus, 6.7 t fresh weight of hay represents 335 small bales of hay per ha.

The number of bales of haylage or hay made per field is usually counted. Reference to Table 2.6 will act as an approximate guide as to whether yield could be improved for a particular site class. Comparisons are often made between one's own yield and that of a neighbour, but if the site classes are dissimilar then yields are going to be different, and the differences are not necessarily due to bad management.

In a study of the nutrient content of silage in Switzerland, Wyss (2000) showed that using fertiliser increased the crude protein and fibre level but decreased the sugar level. This led to the conventional silage scoring lower for fermentation quality.

References

Alberda, Th. (1965) Responses of Grasses to Temperature and Light. In: *The Growth of Cereals and Grasses* (eds F.L. Milthorpe & J.D. Ivins), pp. 202–12. Butterworths, London.

Crompton, E. (1958) Hill soils and their production potential. *Journal of the British Grassland Society*, **13**, 229–37.

Curll, M.L., Wilkins, R.J., Snaydon, R.W. & Shanmugalingham, V.S. (1985) The effects of stocking rate and nitrogen fertilizer on a perennial ryegrass – white clover sward. *Grass and Forage Science*, **40**, 129–40.

Evans, T.W. (1951) *Land Potential*. Faber & Faber, London.

Foster, C. & Lampkin, N. (1999) *Organic Farming in Europe: Economics and Policy* Vol. 3. Univ of Hohenheim.

Frame, J. (1992) *Improved Grassland Management*. Farming Press, Ipswich.

Halberg, N. & Kristensen, I.S. (1997) Expected Crop Yield Loss when converting to Organic Dairy Farming in Denmark. *Biological Agriculture and Horticulture*, **14**, 25–41.

Harrod, T.R. (1979) Soil Survey Applications (eds M.G. Jarvis & D. Mackney). *Technical Monograph 13*.

Hogh-Jensen, H. (2000) Clovers in legume-based grassland maintain their ability for fixing atmospheric nitrogen under low phosphorus and potassium conditions but change the herbage quality. *Proceedings of the 13th International IFOAM Scientific Conference*, Basil, Switzerland, p. 91.

Hunter, R.F. & Grant, S.A. (1971) The effect of altitude on grass growth in East Scotland. *Journal of Applied Ecology*, **8**, 1–19.

Hutchinson, A. (1982) Big bale silage – a practical alternative? *Farm Marketing Services Report 30*, Milk Marketing Board, Reading.

King, F.H. (1926) *Farmers of Forty Centuries*. Jonathan Cape Ltd, London.

Lazenby, A. & Doyle, C.J. (1981) *Grassland in the British economy – some problems, possibilities and speculations* (ed. J.L. Jollans). Grassland in the British Economy, Centre for Agricultural Strategy Paper 10, University of Reading, Reading.

Loges, R., Kaske, A., Ingeversen, K. & Taube, F. (2000) Yield, forage quality, residue nitrogen and nitrogen fixation of different forage legumes. *Proceedings of the 13th International IFOAM Scientific Conference*, Basil, Switzerland, p. 83.

Men, F.P., Urs, Z., Tschachtli, R. & Dubois, D. (2000) Inputs, yields and economic parameters of three farming systems compared at Burgrain (Switzerland). *Proceedings of the 13th International IFOAM Scientific Conference*, Basil, Switzerland, pp. 386–9.

Newton, J.E. (1992) *Herbage production from organic farms.* Research Note **8**, Elm Farm Research Centre, Newbury.

Newton, J.E. (1993) *Organic Grassland.* Chalcombe Publications, Maidenhead.

Peel, S.J., Matkin, E.A. & Huckle, C.A. (1988) Herbage growth and utilized output from grassland on dairy farms in south-west England: Case studies of five farms, 1982 and 1983. *Grass and Forage Science*, **43**, 71–78.

Thomas, C. & Young, J.W. (1981, 1992) *Milk From Grass.* ICI and Grassland Research Institute, Hurley.

Wyss, U. (2000) The influence of different farming systems on nutrient contents and on grass silage quality. *Proceedings of the 13th International IFOAM Scientific Conference*, Basil, Switzerland, p. 398.

3 Dairy, Beef and Sheep Production

There are many satisfactory books describing conventional animal production systems. The purpose of this chapter is not to duplicate any of this information but to propose achievable targets for organic systems of animal production and to comment on reasons for any differences between conventional and organic.

Milk production

In a report commissioned by the South West Regional Development Agency and the Countryside Agency in 2002, it was forecast that organic production is set to double by 2005 and that organically managed land will increase by 100%.

In 1992, Redman noted that British Organic Milk Producers had 45 members, and that average herd size was 73 cows, producing 5384 litres of milk per cow, with a stocking rate at grass of 1.79 cows per ha.* Multiplied up, this suggests that there were 3285 organic dairy cows, producing 17.7 million litres of organic milk per annum, and grazing 1835 ha of grassland.

Figures for the numbers of organic dairy cows in Europe (Foster & Lampkin, 1999) indicated that in 1996 Great Britain still had only 3436 organic dairy cows, compared to Austria (87 068), Switzerland (32 504) and Denmark (21 417). By 1999 Padel, *et al.* (2000) calculated that there were 45 million litres of organic milk produced in the UK. From 2000 to 2003 organic milk production in the UK increased by approximately 600%, and the total organic milk collected in 2003 was 293.4 million litres, of which 62% was sold as 'organic' (Soil Association, 2003).

---

* This had increased to over 350 farms by 2002; see Chapter 10, section on Yeo Valley Organic Company, for details of developments.

Performance figures for organic and conventional herds are shown in Table 3.1. The milk yield figures are slightly lower for organic herds, but there may well be a difference in the breed composition of the respective herds, and it must be remembered that Jerseys and Guernseys give only 75% as much milk as Friesians. In a study of 7 organic dairy herds versus 111 conventional dairy herds in Canada, Stonehouse, *et al.* (2001) recorded a milk yield of 5777 litres per organic cow compared to 6030 litres per conventional cow.

Table 3.1 Performance figures for organic and conventional herds.

	Conventional	Organic
Milk yield (l/cow)	5838	5384
Concentrate use (kg/cow)	1462	1145
Concentrate use (kg/l)	4.0	4.7
Stocking rate (cows/ha)	2.22	1.79

	Good herd	Poor herd
Replacement rate (%)	18	22
Calving to conception (days)	81	107

(Redman, 1992; Webster, 1993)

The milk yield of the top 25% of herds has increased with time: in 1981–2 (Wilkinson, 1984), yield was 5748 litres/cow and by 1992 the yield for the top 25% was 5838. Concentrate use and stocking rate at grass are clearly related. Thus, the more concentrates fed, the higher the stocking rate at grass is likely to be, because concentrates act as a substitute for grass rather than a supplement. Organic standards (1992) insist that at least 60% of the dry matter in all ruminant diets should be either fresh green food or unmilled forage. Assuming that a Friesian cow (bodyweight 600 kg) will eat 14.2 kg dry matter (forage and concentrates) per day (Amies, 1981) then total annual intake = 5.2 t, of which 1.0 t can be concentrates. Corresponding figures for a Guernsey cow (450 kg) are 4.3 t per annum and a concentrate allowance of 1.0 t per annum. This is the allowed amount of concentrates, but the emphasis should be on the reduction of concentrates and an increase in the use of grassland.

In the conventional herd, concentrate use has decreased and stocking rate has increased, presumably because of the increased use of nitrogen fertiliser (Table 3.2). Wilkinson (1984) presents data to

Table 3.2 Concentrate use and stocking rate in the conventional herd.

	1981–2	1986	1992
Concentrate use (t/cow)	1.76	1.6	1.5
Stocking rate (cows/ha)	2.14	2.16	2.22

(Wilkinson, 1984; Webster, 1993; Redman, 1992)

show that the two most important components in dairy profitability are higher stocking rate (59%) and higher milk yield per cow (35%), based on data from the Milk Marketing Board Farm Management Services. Webster (1993) presents data from the same source, and concludes that the differences in milk yield, stocking rate and concentrate use per cow between the top and bottom 25% of farms surveyed are really very small and that the main difference in physical performance is in replacement rate (18% versus 22%). Clearly, he says, the less successful herds have to cull more cows for infertility, mastitis, lameness and poor performance.

In a Norwegian study of 29 organically managed dairy farms and 87 conventional farms Reksen, *et al.* (1999) found an annual replacement rate of 23% in organic and 35% in conventional herds. This lower replacement rate on organic dairy farms was confirmed by Stonehouse, *et al.* (2001). Newman Turner (1955) strongly contends that an organic system should be judged by the health and reproductive performance of the dairy cows. This point was emphasised by Williams (1991), who pointed out that at the beginning of the twentieth century more dairy cows lasted for ten lactations, not three or four as at present. With the advent of cubicles with concrete floors, foot problems are now one of the main reasons for culling cows. Additionally, there is much greater financial pressure to ensure that the level of milk yield is kept as high as possible. This may well be achieved by the introduction of young stock of higher genetic potential.

Newman Turner (1955) suggests that the three reasons responsible for the large increase in disease in conventional dairy herds are: intensified exploitation, artificial feeding and 'the mad race for higher and higher yields'. There is evidence from Germany that the interval to successful rebreeding is shorter and the incidence of disease, particularly mastitis, is lower in organic than in conventional dairy herds; but a significant difference in a major experiment has not been established as yet. If the average count of white cells in the milk rises above 400 000, then milk price is reduced. Organic dairy farmers must be watchful and use good husbandry to control the cell count level, whereas their conventional neighbours can rely on the frequent use of antibiotics.

Research in organic dairy production is growing. A project at the Scottish Agricultural College lists the desirable traits for the organic dairy cow as 'disease resistance, fertility levels, longevity, liveweight, milk yield and milk quality' (Barrington, 2000). From this list order it can be seen that milk yield is only number five out of six.

Walsh (1982) concluded, after studying top grassland dairy farms:

'... that making full use of grass relies on having faith in its ability to fulfil a more demanding role in dairy herd nutrition. Such faith can only stem from growing grass in sufficient quantity and presenting it to the cow at a satisfactory stage of growth and quality.'

Top conventional grassland dairy farmers (Wilkinson, 1984) used 338 kg of nitrogen per ha. They substituted the fertiliser bag for the concentrate sack but, with the advent of the quota, milk should be produced at the lowest possible input cost. Here legumes, particularly white clover, have a most important contribution to make, and one that has direct appeal to organic farmers. It has been shown that cows grazing white clover ate more dry matter than those offered perennial ryegrass, and during the grazing season averaged 13% more milk per day (22.2 versus 25.0 litres per cow per day). There were only slight differences in milk composition and flavour, but the clover pasture gave rise to milk of higher casein content, which improves cheese-making characteristics.

Grassland utilisation

Peel *et al.* (1988) measured the apparent efficiency of utilisation of grazed grass on five commercial dairy farms, and calculated that utilisation averaged 67%, with a wide range from 51% to 83%. The lowest values were on poorly drained farms. This value was similar to the efficiency of utilisation of herbage conserved (silage rather than hay), which averaged 64%, with a range from 55% to 73%. Not surprisingly, they concluded that there is considerable potential for increasing output from dairy farms: higher grazing pressure and more flexible management are needed. Stocking rates from this trial and target stocking rates based on my own experiment (Chapter 2) are shown in Table 3.3.

Table 3.3 Actual stocking rates and potential stocking rates on dairy farms.

	Conventional[1]	Organic[2]	Target organic
Stocking rate (cows/ha)	2.3	1.79	2.0
Fertiliser N (kg N/ha)	232	0	0

[1] Peel, *et al.*, (1988)
[2] Redman, (1992)

The assumptions underlying the target organic stocking rate are:

● A herbage yield of 13 t dry matter per ha per annum
● Utilisation % of 67
● Annual dry matter consumption by a 600 kg Friesian of 5.2 t, of which 1.0 t are met by concentrates, leaving a herbage requirement of 4.2 t dry matter.

The corresponding target stocking rate for a 450 kg Guernsey is 2.4 cows per ha.

How can utilisation be improved?
Wilkinson (1984) presents information showing that the amount of metabolisable energy utilised on dairy farms can vary by more than a factor of four. He suggests that high levels of efficiency can be achieved by:

● close grazing
● frequent silage cuts
● adequate lime, phosphate and potash
● healthy swards.

The organic farmer must be fully satisfied that close grazing does not result in under-nutrition. It need not, but any system based on utilising herbage as fully as possible tends to make the dairy cows eat second or third choice herbage, which may reduce intake slightly and thus milk yield.

There are two methods of ensuring close grazing with minimal wastage (Fig. 3.1).

Rotational grazing This requires either electric fencing plus a back fence, which is comparatively easy with dairy cows, or the creation of

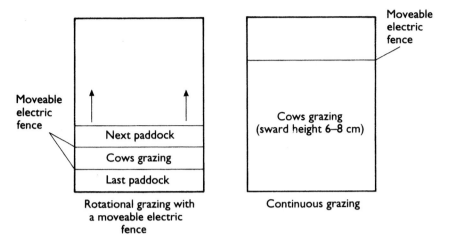

Fig. 3.1 Rotational and continuous grazing.

small paddocks using extra fencing. The advantage of rotational grazing is that it is very easy to see how much herbage is available and it facilitates decisions as to how many areas can be conserved for silage. The disadvantage, particularly in a wet spring, is that very high local stocking rates compound the problem of poaching. The other disadvantage is that intake is likely to vary between days, unless the area allocated for grazing is changed daily. If the paddock is meant to last, or does last, for three days or more, then intake will be highest on the first day, when there is plenty of herbage, and then gradually diminish.

Continuous grazing with the use of the swardstick (or a graduated wellington boot)   The point of measurement is to keep the sward at a height of between 6 cm and 8 cm for dairy cows. If it is higher than this, grass wastage will increase; if it is lower than 6 cm, then intake will fall. The method of keeping sward height within these bounds is to have a moveable electric fence. Measurements need to be taken at least three times a week, particularly during periods of rapid growth in the spring. About a hundred measurements should be made on each occasion in a 5 ha paddock. This is time-consuming, but the better management should result in a dense, leafy sward with minimal wastage.

Conservation

This is of great importance for dairy cows, as the indoor period may well last for more than half the year, and organic silage is almost impossible to purchase. It is still impossible to forecast weather conditions much more than a few days ahead. For this reason, dairy farmers tend to err on the cautious side by not putting up enough paddocks for silage in the spring, in case there is a drought. However, by conserving as many paddocks as possible in the spring and early summer, it ensures that the grazed paddocks are not allowed to accumulate unused patches of senescent herbage, thus improving sward quality; furthermore, silage can be fed back to the cows during the grazing season.

A number of Farm Marketing Services bulletins (Taylor, 1982; Poole & Mabey, 1987; Poole, 1988) highlight increased stocking rate as the major difference between the most and least profitable dairy farms. The assumption is then made that the least successful dairy farmers should increase the efficiency of their grassland utilisation. Whilst this is a worthy aim, the surveys have not actually shown that the difference in stocking rate was not caused by differences in grassland potential yield. No measurement of grassland yield was made. What happened was that the energy requirement for main-tenance, weight change and milk yield of the dairy cow was calculated from performance, then the amount of energy as concentrates was subtracted and the remainder was assumed to be supplied by the herbage, either grazed or as silage. Not only was the energy con-centration of the silage grass assumed, but so was the yield potential of the field. The energy requirement of the cow supplied by grass and silage, divided by the potential yield of the grassland area, was assumed to be the efficiency of utilisation. To repeat, no actual measurement of the grassland yield was made. Whilst it is reasonable to recommend that most farmers can increase the efficiency with which their grassland can be utilised, the differential stocking rate between the dairy farmers with the highest gross margin and those with the lowest may have been caused by differences in grassland potential rather than management. It would have been interesting to swap the managers round!

Stocking rate

Because the rate of grass growth varies during the season, stocking rate will also vary. Overall stocking rate will depend on the grass growing potential of the fields (see Chapter 2), but actual stocking rate will be higher in the spring than later in the year when the rate of herbage growth is beginning to decline. The growth pattern of a grass/ white clover sward is more even than that of ryegrass with bag nitrogen, so that a ratio of 1.5:1 is recommended, with the first part of the season being turnout to mid-July and the second part being mid-

July to housing. For a dairy farm with half site class 2 fields and half site class 3 fields, yielding 12 t dry matter per ha per annum and utilising this with an efficiency of 67%, annual stocking rate would be 1.9 dairy cows per ha. With a spring to autumn stocking rate ratio of 1.5:1, stocking rate from turnout to mid-July would be 2.28 cows per ha and from mid-July to October would be 1.52 cows per ha.

Soil nutrients

Adequate levels of lime, phosphate and potash in the soil are especially important for organic dairy farmers, because these nutrients are essential for healthy white clover. The supply of these nutrients and general manure strategy is covered fully by Newton (1993).

Autumn and spring calving

Autumn calving should occur in the three months of September, October and November, although from survey evidence it is clear that most dairy farmers calve their cows over a longer period, calving occurring from August to May, with only 51% of the herd calving between September and November (Taylor, 1982).

Spring calving occurs in the months of March, April and May, but again a three-month calving period is rarely obtained in practice, nor is it necessarily desirable for organic farmers to block calve, particularly if they are the sole supplier of a specialist market that requires an even supply of milk through the year.

Because the trend towards autumn calving has been accentuated (Poole, 1988), there has been a shortage of milk for making into butter and cheese, particularly in August and September. The Milk Marketing Board (now Milk Marque) therefore increased the payment for milk produced in July, August, September and October. These differentials have lifted the income from cows calving between January and August above that of the autumn calvers.

Cows calving between September and November have the highest requirement of concentrates, whereas February to April calvers have the lowest requirement. Thus spring calvers on organic dairy farms are more likely to meet the forage dietary requirements of the Soil Association regulations than autumn calvers. Furthermore, should the concentrate allowance be cut further, then this time of calving will be increasingly favoured.

In a comparison of reproductive performance between organic and conventional dairy herds in Norway, Reksen, *et al.* (1999) found that organic dairy cows bred less successfully when mated in winter than conventional dairy cows. They attributed this to inadequate levels of concentrate fed and suggested a relaxation of the organic standards for the winter feeding of organic cows.

The advantages for autumn calving, particularly on the dairy/arable farm, are:

- The whole herd will be in milk in spring, so that peak grass growth in April, May and June can be fully utilised.
- The herd will dry or be drying off during the uncertain period for grass growth, centred on August.
- Sound guidelines exist for autumn calving, and it has been seen to be successful and therefore popular.

The advantages of spring/summer calving are:

- Calving in the spring/summer will normally take place in the field, in a maternity paddock. Calving outside tends to give a good clean, disease-free environment, which leads to strong, easy-to-rear calves and a good calf price for those sold.
- Experience shows that cows are more easily got in calf during the summer.
- Heifer rearing is easier with a summer calving herd. Calves are usually strong and free of pneumonia, but the advantage with outdoor calving depends on the cleanliness and suitability of the available housing.
- No problem with summer mastitis.
- Higher milk price.

Beef production

'The impression that beef herd management is a simple business is far from the truth. It confuses the low level of supervision of cows (easy care) with the high standards of organisation and management needed for profitable production. If anything, suckler herds are more complicated to manage than beef finishing enterprises because reproduction, lactation and growth all contribute to the output of the enterprise.' (Allen, 1990)

Suckler herds dominate organic beef production, partly because organic rules forbid the purchase of calves from the market, thus limiting the number of bought-in calves available and partly because feed-lot finishing based on concentrates and indoor housing are also banned.

Factors governing success
In analysing the percentage contribution to extra gross margin per ha of top suckler herds and top 18-month beef producers from data collected by the Meat and Livestock Commission (MLC), Wilkinson (1984) lists the factors that govern success in Table 3.4. He also notes that the cows of top beef producers were back in calf ten days earlier than the average, and the calving period was three weeks shorter.

Table 3.4 Factors governing success.

	Suckler beef	18-month beef
Higher stocking rate (%)	32	42
Higher sale weight	23	10
Higher weaning %	17	—
Lower concentrate use per head	8	22
Lower herd replacement costs	5	—

(Wilkinson, 1984)

A higher stocking rate is as important for success in beef production as in dairy production. But the use of a higher stocking rate may easily be the result of better land rather than better management. There is always scope for better grassland utilisation but, as with sheep, it must be remembered that many beef suckler enterprises are based on relatively unproductive land that does not have many alternative agricultural uses. So those beef enterprises that are based on difficult steep land should not be compared with beef enterprises in the lowlands. These upland areas would be unused were it not for herds of beef cows and flocks of sheep, and beef is being produced with a relatively low input of cereals, protein and fossil fuels.

Grassland management
Nitrogen use on conventional beef farms is much lower than on dairy farms, and there is, therefore, a greater likelihood that stocking rates on grass/white clover swards will match those on grass plus 150 kg of nitrogen per ha. Clover also has the great virtue of promoting faster growth rates and higher milk yields than grass, mainly because of

higher intake (Fig. 3.2). For example, from Fig. 3.2 it can be seen that one animal fed *ad libitum* white clover grew at 1100 g/day while its counterpart grew only 580 g/day on *ad libitum* grass. What is certain, however, is that at pasture the provision of an adequate daily allowance of high quality herbage throughout the grazing season is now recognised as a major challenge to the farmer who wishes to make more money.

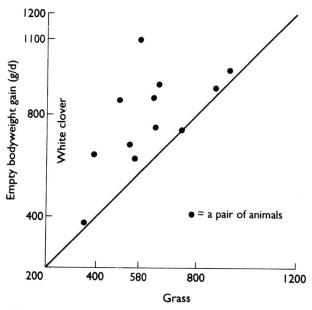

Empty bodyweight gain = liveweight gain corrected for empty bodyweight on the basis of gut-fill values.

Fig. 3.2 A comparison between white clover and grass for fattening beef cattle (Thomson, 1984).

As Wilkinson (1984) states, flexibility in grassland management is essential, particularly when young beef cattle are reared at pasture, or when store cattle or suckled calves are finished on grass. The overall goal must be to improve the predictability of beef cattle growth, so that plans made at the outset for the beef system are realistic and are reflected in the actual level of performance.

Suckler beef production
The key planning decisions are the season of calving and whether calves are sold at the autumn sales, overwintered for sale as stores in

the spring or fed through to slaughter. The intended disposal of calves interacts with the calving season.

The performance and inputs for different times of calving are shown in Table 3.5. Autumn-born calves are older and heavier at the autumn sales than those born in the spring and give a higher gross margin per cow, but spring calving herds have a higher gross margin per ha because they can be stocked more heavily. With winter calving, the calf is quite well grown by the spring and consequently will make good use of grass. The down side is that calving has to be indoors, which is likely to increase the losses from disease. With summer calving, the calf is well grown in the autumn, and winter feeding can be aimed at the calf rather than the cow.

Table 3.5 Effect of time of calving on inputs and performance.

	Calving season			
	Spring	Summer	Autumn	Winter
Calf age at sale (months)	7	15	12	9
Calf daily gain (kg)	1.0	0.8	0.9	1.0
Calf weight at sale (kg)	265	425	375	325
Concentrates (kg) – cows	100	200	150	125
– calves	75	200	125	100
Silage (t) fresh weight per cow	4.5	6.5	6.0	5.0
Stocking rate (cows/ha)	1.75	1.35	1.5	1.65

(Allen, 1990)

On difficult upland and hill farms the scarcity of winter feed dictates spring calving. Similarly, on arable farms the use of crop residues of moderate quality fits in best with spring calving. If there is a shortage of buildings for winter housing, then spring calving is again to be preferred. On good land with plenty of high-quality silage, labour and adequate building space, autumn calving is sensible, and the suckled yearling calves can go straight into winter finishing. Larger farms may even have complementary spring and autumn calving herds to utilise all resources to the full.

Breeds for suckler beef production
No breed or crossbred is 'best' for the wide range of beef production systems. Variations in growth rate and earliness of maturity make

breeds suitable for different systems. The interaction between earliness of maturity and diet quality is the most important of all.

Early maturing breeds such as Angus and Hereford were developed to cope with prolonged and severe winter store periods and to produce beef from low quality feeds. On these feeds, late maturing Charolais crosses would grow bone and some muscle but would not finish satisfactorily. On the other hand, high quality rations which exploit the rapid, lean growth of Charolais crosses would cause Angus and Hereford crosses to become over-fat at relatively light weights. The use of different breeds of dairy–beef crosses in a grass finishing system is shown in Table 3.6.

Table 3.6 Grass finishing of dairy-bred steers.

	Angus × Friesian	Hereford × Friesian	Friesian/ Holstein	Limousin × Friesian	Charolais × Friesian
Daily gain (kg)	1.0	1.0	1.0	1.0	1.1
Grass finishing period (days)	110	120	170	190	170
Slaughter weight (kg)	450	475	550	575	610
Grazing stocking rate (cattle/ha)	5.1	4.8	4.3	4.2	3.9

(Allen, 1990)

The short grazing period of early maturing Angus and Hereford × Friesians is an advantage, because slaughter commences in mid-grazing season (June/July). This reduces cattle numbers and grazing pressure when grass growth rate and digestibility are beginning to decline. By taking the lighter weight of the Angus and Hereford crosses into account, they can be stocked more heavily than the Limousin and Charolais crosses, thus equating liveweight gain per ha.

With the later maturing breeds, concentrates will have to be fed at grass in the autumn, and even then some will not fatten while grazing and will have to be brought in for winter feeding or for another store period. The shorter the growing season, because of latitude or altitude, the better the early maturing breeds perform.

With the grass finishing of suckler herd cattle, Hereford crosses are so early maturing (Table 3.7) that a second batch needs to be organised so that all the grass grown can be utilised. Cattle in the second batch will grow less well because of herbage quality. The longer grazing period for the Charolais cross utilises all the grass grown,

Table 3.7 Grass finishing of suckler-bred stores.

	Hereford ×		Charolais ×
	1st batch	2nd batch	
Daily gain (kg)	1.0	0.9	1.1
Grass finishing period (days)	100	90	165
Slaughter weight (kg)	475	450	550
Grazing gain (kg/ha)		775	775

(Allen, 1990)

while at the same time stocking rate is being reduced in the second half of the year, as finished cattle are sold off.

Finally, it is important to realise that suckler beef is top of the meat trade range and commands the premium prices. Carcasses are of good shape, meat yields are higher than for dairy beef and the eating quality is more highly regarded.

Meili (2002) confirms this from Switzerland. Organic beef sells very well in the biggest supermarket. The consumer pays 10% to 15% more for the meat and the farmer receives 40% more for the organic animal than for the conventional. The favoured cross for quality organic beef is the Limousin crossed with a dairy breed. This cross can be slaughtered at 550 kg liveweight to make a 300 kg carcass.

Body condition score and calving interval
It was pointed out earlier that the cows in the top beef herds have a shorter calving interval and a tighter calving pattern; both are influenced by body condition score (bcs), which is easily assessed. The relationship between bcs and calving interval is quite clear. Cows of bcs over 2 and up to 4 have a calving interval of 365 days; below this bcs the interval lengthens and profit drops until at bcs of 0.5 the interval is 450 days.

The calf
Under the Soil Association Organic Food and Farming Standards (1992), it is prohibited to buy organic calves from livestock markets, nor can beef calves from non-organic herds ever count as organic animals. Bucket or artificial-teat rearing on organic whole or reconstituted milk is permitted, and so is non-organic milk or milk replacer free from antibiotics and additives – but only in an emergency. Colostrum is vital, and in case a cow dies or lactation is delayed it is

Table 3.8 Stocking rate and beef suckler enterprise.

	Cows/ha
Lowland herds – autumn calving	1.7
Lowland herds – spring calving	2.1
Upland herds – autumn calving	1.3
Upland herds – spring calving	1.6

(Allen, 1990)

worth keeping a supply of deep-frozen colostrum. Milk yield of the suckler cow is affected by the suckling power of the calf, which is dependent on its birthweight. On average, for each 1 kg increase in birthweight, weaning weight increases by 6 kg.

Targets
Grassland management has a major effect on cattle performance through the quality of grazing and conserved forage, and on gross margin per ha via stocking rate. Stocking rates to aim at for each type of suckler enterprise are shown in Table 3.8. Target stocking weight should be 2.5 t per ha (cow plus calf weight) in the spring and 1.5 t per ha in late season (i.e. after July). The control of parasitic worms is much less of a problem in suckled calf production than with dairy-bred cattle, stocked heavily. Targets for suckled calf production in lowland and upland herds are shown in Table 3.9.

In spring calving herds, it pays to retain calves for sale as stores in the following spring rather than take them to the autumn sale, mainly because of the higher sale price per kg in the spring.

Table 3.9 Targets for suckled calf production in lowland and upland herds.

	Continental sire breeds		British sire breeds	
	Autumn	Spring	Autumn	Spring
Calving period (weeks)	12	8	12	8
Calves reared per 100 cows bulled	92	92	95	95
Calf daily gain (kg)	1.0	1.1	0.9	1.0
Rearing period (months)	10	7	10	7

(Allen, 1990)

Replacement heifers

Replacement rates in suckler herds are typically 15% to 18%, with barrenness and late calving as the main reasons for culling. On lowland and upland farms it is best to calve British breeds at 24 months and continental crosses at 30 months, as they are later maturing. Key production targets are weight and condition score at mating and calving (Table 3.10) for dairy-bred replacements.

Table 3.10 Target weights and condition scores for dairy-bred calves reared as suckler herd replacements.

	Mating		Post calving	
	Liveweight (kg)	Condition score	Liveweight (kg)	Condition score
Spring calving at 2 years (summer mating)				
Angus/Hereford × Friesian	375	3	475	2.5
Limousin/Simmental × Friesian	400	3	525	2.5
Autumn calving at 2 years (winter mating)				
Angus/Hereford × Friesian	350	2.5	500	3
Limousin/Simmental × Friesian	375	2.5	500	3
Spring calving at 2.5 years (summer mating)				
Angus/Hereford × Friesian	450	3	525	2.5
Limousin/Simmental × Friesian	500	3	575	2.5

(Allen, 1990)

Good grazing gains are important and the leader/follower system is ideal, with the younger heifers grazing ahead of the older generation. Rearing costs are minimised by holding gains at modest store levels during the winter and then anticipating compensatory growth during the grazing season. Heifers should be mated to calve early in the calving period, to allow for delayed rebreeding. After calving on winter diets, extra feed should be offered, as the heifers are still growing as well as suckling their calves (Table 3.11).

Disease

Parasites

Beef calves can suffer serious problems in health and performance if care is not taken to control intestinal parasites (*Ostertagia* spp. par-

Table 3.11 Targets for rearing autumn-born calves, calving at two years old.

	Angus/Hereford × Friesian	Limousin/Simmental × Friesian
Reared calf (kg)	95	105
Daily gain (kg)		
rearing winter	0.6	0.6
first summer	0.7	0.8
second winter	0.6	0.6
second summer	0.7	0.8
Concentrates (kg)	300	500
Silage (t, 25% dry matter)	5.5	6.0
Stocking rate (cattle/ha)	1.7	1.6

(Allen, 1990)

ticularly). This is best done by a clean grazing system, alternating beef, sheep and conservation, if possible, in consecutive years.

Pneumonia
Good ventilation in buildings is essential to minimise the incidence of pneumonia. Buildings that seem exposed by human standards are much healthier for cattle than those with a stuffy atmosphere. Stale, humid air, laden with dust, is a recipe for enzootic pneumonia.

Lungworm
This is recognised as a problem, and the recommended procedure from the Organic Food and Farming Standards (1992) is to allow suckled calves to develop natural immunity by grazing grass with their mothers. If there is a known farm problem, then oral husk vaccine can be used.

Sheep production

The European country recorded with the most organic sheep in 1996 was Austria (99 275), followed by Great Britain (36 231), Sweden (26 652) and Switzerland (19 610) (Foster & Lampkin, 1999).

Benoit, writing in 2002, felt that organic sheep have a profitable future in France. He calculated that with a 20% to 30% higher price for organic lamb and by extensifying the area of forage crops so that the farm is self-sufficient for food, then farm income can be maintained or even increased.

In Iceland, Dyrmundsson (2002) stated that there is likely to be a growth in organic sheep products due to the strong environmental and quality image of lamb and wool. Moreover, the positive links between sheep farming and both environmental management and the maintenance of rural communities could counteract the trend towards the industrialisation of agriculture. However, in most Nordic countries organic lamb has to compete with the high reputation of conventional lamb.

Systems of sheep production

Sheep are seasonal breeders so, with the exception of breeds such as the Dorset Horn, lambing is restricted to the period between February and May. The short breeding season dictates the supply of lamb and makes for an uneven supply of finished lamb for the market. The months of shortage are February to June.

Early lambing

This can be achieved naturally from November onwards, using the Dorset Horn or Dorset Horn crosses. The aim of the system is to produce saleable lambs for Easter, when prices are at their highest. Because the ewes are being asked to reproduce at the very start of the breeding season, lambing percentages tend to be low: either the number of barren ewes is high, if the rams are left in for only six weeks, or else with a longer period of ram inclusion then the lambing season becomes more protracted. This makes more work for the shepherd, and complicates the rearing and growing of lambs for Easter because of the wider age spread.

From the organic standpoint, early lambing is the least 'natural', partly because of the length of the housing period for the lambs and partly because the easiest and most reliable way to guarantee lamb growth indoors is to feed either concentrates or dried grass pellets. These are expensive and can only be fed to a certain level in the ration, as defined in the Soil Association Standards.

The skill comes in achieving a reasonable selling percentage (150 lambs sold per 100 ewes to the ram), combined with a positive margin over feed costs. Because the lambs are either sold or weaned by the time grass begins to grow appreciably in late March or April depending on latitude, the dry ewes can be kept at high stocking rates at pasture.

Mid-lambing period

March/April is the natural time to lamb, with fresh and growing grass available to the newly lambed ewe. Reproductive performance should be at its highest, with ewes being mated at the second or third oestrus in the breeding season. A compact mating period will result in a short lambing season, with fewer than 3% of the ewes being barren.

The number of lambs born per ewe depends on the breed of ewe and on body condition score at mating. The optimal number of lambs born per ewe depends on the system of production on the farm. For instance, ewes lambing in an upland or mountain environment may well be unable to fatten more than a single lamb adequately, twin lambs being too small to survive the rigours of the mountains, and the ewe unable to find sufficient nourishment to milk well.

On the other hand, lowland ewes should be able to rear 1.75 lambs. This is well within the capability of most crossbred ewes, such as the Masham, Greyface and Scottish Half-Bred. Of the pure breeds, the Lleyn is prolific enough to sell 175 lambs for every 100 ewes to the ram: a figure that allows for barrenness and lamb deaths. The main rearing problem is that to achieve a lambing percentage of 185 (lambing percentage = lambs born alive per 100 ewes to the ram), a quarter of the flock will give birth to triplets. These are difficult to rear on the ewe, so cross-fostering at birth is the most effective solution, but this can be done smoothly only in large flocks, because the foster mother has to be available at the required moment.

Another important aspect of breeding performance is ewe size, which must be taken into account when choosing a stocking rate (Fig. 3.3). Ewes eat according to their weight, and within the multiplicity of sheep breeds and crosses in Britain there is a weight difference that can be fourfold, with the adult Soay weighing 20 kg and the Oxford Down weighing 80 kg. The most efficient nutritional combination is to have a prolific small ewe mated to a large ram, and the least efficient animal is the large ewe with a single lamb. In Britain, the weight and quality of the wool clip, although a useful adjunct, does not compensate financially for poor reproductive performance and expensive nutritional demand.

Lambs born in March and April have the whole of the grass and clover growing season in which to reach target slaughter weight. The main problems are intestinal parasites and low lamb prices during August and September, the time when the greatest proportion of lambs are slaughtered.

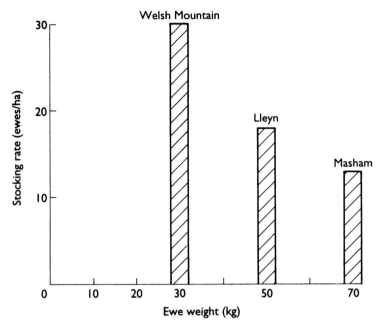

Fig. 3.3 Ewe weight and stocking rate.

Late lambing

With a flock lambing in May, there is very little need to feed any concentrates. The times of greatest nutritional demand – late pregnancy and early lactation – occur in late March, April and early May, when in most years the grass has begun to grow strongly and digestibility is at its highest, because of the high proportion of young leaves. Reproductive performance in May should be as good as in March/April.

The drawback with late lambing is its dependence either on a buoyant store lamb market in September and October or else on the feasibility of growing a catch crop, such as stubble turnips or rape, on the farm, so that lambs can be fattened satisfactorily outdoors during the autumn and early winter. Again, intestinal parasites can be a problem, with the period of maximum risk, late July and August, occurring before the younger lambs have had a chance to build up their own immunity.

Store lamb fattening

Some dairy and arable farms buy in store lambs in the late summer or autumn to utilise excess grass and arable by-products, during the winter. This is a convenient and short-term enterprise that depends for success on buying healthy organic stock from organic farms, at reasonable prices, fattening the lambs and selling them when prices are high in the spring. It is not permissible to buy in conventional store lambs, keep them on symbol grassland and then sell them as organic.

Performance and output targets for the three main production systems are shown in Table 3.12. Selling percentage, which is the crucial output figure for reproductive performance, takes into account ewe barrenness and lamb mortality. Thus to achieve a selling percentage of 175 it will probably be necessary to have a mean litter size (number of lambs born per ewe lambing) of 2.0. Barrenness in mid and late lambing flocks should be below 3%, and lamb mortality from birth to sale should be below 10%. The growth rate of the singles and twins will be slower for the late lambing system because their time at grass is shorter.

Table 3.12 Performance targets for different times of lambing.

| | | Lamb growth rate (g/day) | |
	Selling percentage	Singles	Twins
Early lambing	150	300	250
Mid lambing			
medium ewe	150	300	250
large ewe	175	300	250
Late lambing			
medium ewe	150	250	200
large ewe	175	250	200

(Newton, *et al.*, 1982)

The effect of ewe size and reproductive performance on ewe stocking rate and lamb output per ha is shown in Table 3.13. It has been assumed that grass output is 10 t of dry matter per ha per annum and that this is used with an efficiency of 67%. The small ewe plus 1 lamb eats 300 kg of dry matter per annum, the medium ewe eats 500 kg and the large ewe eats 600 kg. It is also assumed that the lamb from the small ewe will weigh 34 kg, from the medium ewe 38 kg and from the large ewe 40 kg. Although the small ewe eats less, the higher stocking rate does not fully compensate for the lower reproductive performance, and this is shown in the lower output of lamb meat per ha.

Table 3.13 Effect of ewe size and reproductive performance on ewe stocking rate and lamb output per ha.

	Ewe stocking rate (per ha)	Lamb output (kg/ha)
Small ewe (30 kg) plus 1 lamb	20	680
Medium ewe (50 kg) plus 1.5 lambs	13	740
Large ewe (70 kg) plus 1.75 lambs	11	770

(Newton, *et al.*, 1982)

Stocking rate has been shown per annum, and allows for hay or silage for the winter to be made during the grass growing season from the land allocated for sheep. In the early lambing system, much more hay or silage needs to be made because the ewes and lambs are likely to be housed for four months, and the lambs will also be eating hay or silage. Overall stocking rate is likely to be the same, but during the grass growing season a greater proportion of the land will be conserved, and the stocking rate for the dry ewes will be proportionately higher than for the mid and late lambing systems. Once the lambs are weaned, their stocking rate will be high because they are smaller than the ewes, except in the case of the 30 kg ewes.

Organic wool
There are now agreed standards for organic textiles, so organic wool can be made into organic products.

The processing and manufacture of textiles involves grading, cleaning, washing, bleaching, dyeing and finishing, and then there is still weaving and knitting.

In Britain, all wool, with the exception of 12 rare breed wools, has to be sold through the British Wool Marketing Board – that is for all sheep owners with more than four sheep. The co-operation and agreement of the BWMB as to organic regulations for wool and its processing is therefore essential. Other countries, Australia, Germany, the Netherlands and Sweden, plus Texas in the USA, already have standards for organic textiles and these have helped in framing British regulations.

What is certain is that the wool clip has very little value to the farmer at the moment. In the case of many breeds, the fleece fetches less than the cost of shearing. It is to be hoped that those sheep farmers who become organic will eventually receive extra for their wool.

Efforts are currently being made to expand the uses of wool for such products as baby goods, felt cushions, shoes, tents and for building insulation. There is undoubtedly a demand in Europe for wool free of organo-phosphates. See Chapter 10, under the heading *Good Herdsmen*.

Grassland management

As with dairy and beef systems, higher stocking rate has been identified as making the highest percentage contribution to extra gross margin per ha in both lowland (38%) and upland (68%) flocks (Wilkinson, 1984). Undoubtedly, grazing management can be improved on most sheep farms, but, as mentioned in earlier sections, the survey is most likely picking out better land as well as better managers.

The use of white clover is of great importance for organic sheep farmers, partly because it boosts herbage yield by fixing nitrogen and partly because the presence of plenty of clover in the nutrition of lambs can increase growth rate by 25%, compared with grass on its own (Fig. 3.4). The problem with sheep is that, unlike cattle, they select clover rather than grass and, furthermore, because of the way they graze they can eat clover without eating grass at the same time. They harvest herbage with their teeth, rather than wrapping their tongue round a clump of material and pulling like cattle.

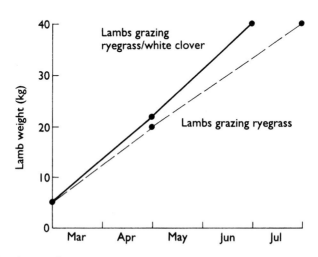

Fig. 3.4 Lamb growth rate on ryegrass or ryegrass/clover.

To prevent white clover being eaten out of the sward by sheep, it is preferable to rest the field periodically. Ideally, in a grazing system with sheep and lambs, rotational grazing with forward creep for the lambs should be practised. In a six paddock system, each paddock can be grazed for five or six days and then the animals are moved on. The lambs get the pick of the grazing ahead of their mothers and therefore grow faster, while with the rest period, which should be about 30 days, the clover is given a chance to recuperate. Length of rest period will vary with speed of herbage growth. During the periods of fastest growth, the sheep will stay in each paddock for longer; but to prevent the herbage ahead of the animals becoming too coarse, paddocks can be taken out for conservation. This is the best grazing method for ewes and lambs, but one that can be practised only if all the grassland for the sheep is in one block. If these were dairy cattle, then trouble would be taken but, because they are sheep, less money will be spent on them.

Oestrogens
The only drawback with using red clover for sheep is that it contains oestrogens that reduce reproductive performance, so it is best to avoid its use during the mating period. Cattle render these oestrogens harmless. White clover and lucerne can also be oestrogenic, but only when the leaves are diseased; the oestrogens produced by the disease in the leaves are different from those found in red clover, which are always present. The oestrogens in white clover and lucerne are also different from each other. Oestrogenic lucerne affects reproduction in sheep, but oestrogenic white clover appears to have no effect.

Health problems with sheep

Internal parasites
The most satisfactory method for the organic farmer to reduce the risk of infection to lambs from internal parasites is to practise clean grazing, which reduces parasitic infection and increases lamb growth rate (Fig. 3.5). The most effective method of clean grazing is, as we have seen, to use a three year rotational system with sheep, followed by cattle, and then arable. This presupposes land that can be ploughed. On permanent pasture farms with no arable, beef and sheep should be alternated. To make this effective, it helps if there are as many beef livestock units as sheep. If the farm contains only sheep, then it is advisable to alternate on an annual basis between ewes with twins and ewes with singles. If the flock normally produces mostly

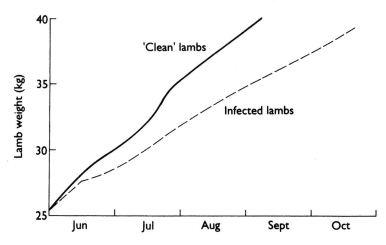

Fig. 3.5 The growth rate of lambs grazing 'clean' or infected pasture (Coop, 1986).

singles then the problem of parasites should not be too acute, as singles are less vulnerable to parasitic infection than twins because a higher proportion of their diet is milk. Internal parasites continue to be of concern to organic farmers (Keatinge, 1999; Pattison, 2001).

In Hawaii, Faust (2000) contended that the hair sheep, St Croix breed, not only has a unique resistance to parasitic worms but also keeps the coffee, avocado and citrus orchards free of tropical weeds.

Fly strike and scab
The Soil Association prohibits the use of organo-phosphate dips against fly strike and scab, mainly because of the undoubted health risk to shepherds. However, there are dips that can be used which are effective and which do not contain organo-phosphates. Flumethrin-based compounds can be used against sheep scab.

It should be emphasised that if animals are suffering then the farmer is encouraged to use the most effective veterinary method for alleviating suffering as quickly as possible. However, attention to husbandry, nutrition and stress factors will markedly reduce the incidence of disease.

References

Allen, D.M. (1990) *Planned Beef Production and Marketing*. Blackwell Science, Oxford.

Amies, S.J. (1981) Block calving the dairy herd. *Farm Management Services Report 26*, Milk Marketing Board, Reading.

Barrington, L. (2000) Breeding for Organic Dairy Systems. *Organic Farming*, **68**, 23.

Benoit, M. (2002) Organic Meat Sheep farming in France: Which farms and what results? Importance of food self-sufficiency. *Proceedings of the 14th IFOAM Organic World Congress*, Victoria BC, Canada, p. 89.

Coop, R. (1986) Subclinical parasitism: The insidious effects of roundworms on lamb performance. *Science and Quality Lamb Production*, 24–5, Agricultural and Food Research Council, London.

Dyrmundsson, O.R. (2002) Organic Sheep Farming under Nordic Conditions–Iceland. *Proceedings of the 14th IFOAM Organic World Congress*, Victoria BC, Canada, p. 91.

Faust, R.H. (2000) The use of Hair Sheep in Organic/Poly-Culture Tree Crop Weed Management. *Proceedings of the 13th International IFOAM Scientific Conference*, Basel, Switzerland, p. 425.

Foster, C. & Lampkin, N. (1999) *Organic Farming in Europe: Economics and Policy*, Vol. 3, University of Hohenheim, Germany.

Keatinge, R. (1999) Controlling the roundworm. *Organic Farming*, **61**, 20–22.

Meili, E.A. (2002) Low Input Milk and Beef Production in Organic Farming. *Proceedings of the 14th IFOAM Organic World Congress*, Victoria BC, Canada, p. 88.

Newman Turner (1955) *Fertility Pastures*. Faber & Faber, London.

Newton, J.E. (1993) *Organic grassland*. Chalcombe Publications, Maidenhead.

Newton, J.E., Betts, J.E., Orr, R.J., Wilde, R.M. & Dhanoa, M.S. (1982). The effect of time of lambing on sheep production. *Technical Report 30*, Grassland Research Institute, Hurley.

Padel, S., Fowler, S., McCalman, H. & Lampkin, N. (2000) Financial implications of organic dairy production in the UK. *Proceedings of the 13th International IFOAM Scientific Conference*, Basel, Switzerland, p. 670.

Pattison, R. (2001) Internal Parasites. *Organic Farming*, **69**, 24–29.

Peel, S.J., Matkin, E.A. & Huckle, C.A. (1988) Herbage growth and utilized output from grassland on dairy farms in south-west England: case studies of five farms, 1982 and 1983. *Grass and Forage Science*, **43**, 71–8.

Poole, A.H. (1988) Response to changes in the seasonality payments. *Farm Marketing Services Report 61*, Milk Marketing Board, Reading.

Poole, A.H. & Mabey, S.J. (1987). An Analysis of Farm Management Services Costed Mixed Farms 1986–7. *Farm Marketing Services Report 58*, Milk Marketing Board, Reading.

Redman, M. (1992) Organic Dairy Costings. *New Farmer and Grower*, **33**, 20–22.

Reksen, O., Tverdal, A. & Ropstad, E. (1999) A comparative study of reproductive performance in organic and conventional dairy husbandry. *Journal of Dairy Science*, **82**, 2605–2610.

Soil Association (2003) *Organic food and farming report*. Soil Association, Bristol.

Soil Association Standards (1992) Revision 5, Soil Association, Bristol.

Stonehouse, D.P., Clark, E.A. & Ogini, Y.A. (2001) Organic and conventional dairy farm comparisons in Ontario. *Biological Agriculture and Horticulture*, **19**, 115–125.

Taylor, K. (1982) Systems of Dairy Farming. 1. The Specialist Grass Farm. *Farm Marketing Services Report 32*, Milk Marketing Board, Reading.

Thomson, D.J. (1984) The Nutrititive Value of White Clover. *Forage Legumes, Occasional Symposium 16*, British Grassland Society, 78–92.

Walsh, A. (1982) The contribution of grass to profitable milk production. *Rex Paterson Memorial Study*, British Grassland Society, December 1982.

Webster, J. (1993) *Understanding the Dairy Cow*, 2nd edn. Blackwell Science, Oxford.

Wilkinson, J.M. (1984) *Milk and Meat From Grass*. Granada Technical Books, London.

Williams, D. (1991) *The Healthy Rearing of Livestock*. British Organic Farmers and Organic Growers Association (BOF/OGA) 7th National Conference, Cirencester.

4 Pig and Poultry Production

Pig production

Both in Germany (Klumpp & Haring, 2002) and Austria (Gruber, *et al.*, 2000), there is a growing and, as yet, unsatisfied demand for organic pork. Farmers thinking of converting to organic pig production cite high labour requirements for roughage production and feeding, and a sub-optimal fattening efficiency due to a diet lacking in adequate protein for the small numbers of organic pig producers. In 1996, the top four European countries for numbers of organic pigs were Austria (39 820), Switzerland (23 643), Germany (23 737) and Denmark (19 550), with Great Britain having 7203 (Foster & Lampkin, 1999).

Pigs are very efficient converters of cereals compared with ruminants, and this efficiency is enhanced by temperature control and minimal exercise. So the pig has changed from being a converter of waste products to being a producer of cheap meat based on cereals. Pig production, unlike dairying, beef or sheep production, is an unsupported industry with no grants, although the guarantees are less certain than they used to be and are changing fast. This means that if there is slight oversupply of pigs on the market then the price drops rapidly, and a system that was in profit quickly turns into a loss maker.

For this reason, pig farmers have had to concentrate on being extremely efficient and to incorporate any research finding that would enhance this efficiency, as rapidly as possible, often to the detriment of the welfare and freedom of the pig. The indoor keeping of pigs in confined conditions had to become dependent on drugs to maintain the sort of growth rates and performance that were required to keep the large units in profit. And with a large indoor unit there is a great deal of pig slurry to dispose of. This has led many people to object to meat-eating on moral grounds. They object to 'the indiscriminate use of drugs, careless pollution of the environment and lack of sufficient concern about farm animal welfare' (English, *et al.*, 1988).

The problem with keeping pigs outdoors is particularly acute in the Netherlands (Janmaat & Groeniger, 2000), because the permitted ammonia emission is so limited. They are currently experimenting with organic pigs being housed in a greenhouse-like structure with plenty of fresh air and daylight, which will reduce the amount of ammonia emission. Despite this major obstacle, the number of organic pigs was forecast to rise from 5000 in 1998 to 469 000 in 2002.

Because the organic pig producer is very concerned about welfare and can use no drugs, the pigs have to have more room, so that there is less risk of infection: organic pigs are kept mainly outdoors, as long as the conditions outside are suitable, that is not too hot, cold or wet and muddy. Ironically, because of the higher capital costs of large indoor units, more pig keepers since 1990 are choosing to invest in outdoor pig systems.

Breeds
The breeds of pig that are most suited to living outdoors, such as the black Wessex Saddleback (now known as the Saddleback), are not those with the thinnest backfat. Docility, good mothering instincts, high food intake and resistance to sunburn; all these excellent characteristics for the outdoor pig make the Wessex Saddleback a good choice, but the high food intake produces offspring that are more likely to have P2 measurements (backfat thickness) of 13, 14 and 15 probe, whereas the highest premium is paid for 12 probe. To some extent, using Landrace boars will slim the offspring down.

Some organic pig farmers, who mostly keep a few sows in the orchard, favour the Gloucester Old Spot. This breed produces fatter meat than the Wessex, and the successful sale of this breed needs a sympathetic local butcher with a reliable local demand.

Reproduction
There has been a steady increase in the number of weaners reared conventionally per sow per year (Fig. 4.1) and the average was 21.3 in 1987. Because there are only a few organic pig producers at the moment, it is only possible to quote performance figures from limited data for the average number of organic weaners reared per sow per year, which is 18 (H. Browning, pers. comm.).

The problem for organic producers, as for other pig farmers who have outdoor systems, is that piglet death is highly correlated with

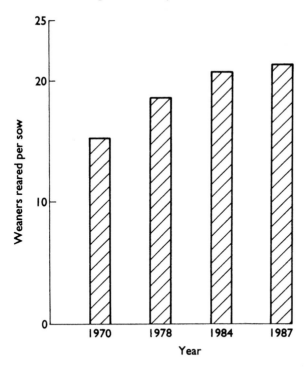

Fig. 4.1 Weaners reared per sow per year (English, *et al.*, 1988).

birthweight. The objective is for the piglets to weigh 1.3 kg at birth. If they are less than 0.8 kg then they have less than a 50% chance of survival. Furthermore, new-born piglets do best at a temperature of 29°C. Lower temperatures decrease survival and growth rate.

The use of farrowing crates is prohibited for organic producers, and so are routine teeth cutting and the automatic use of iron injections. However, a protective rail, farrowing box or nest is recommended, and teeth cutting for individual piglets or a litter when necessary to prevent injury to the sow is permitted, and so are iron injections for anaemia in the case of iron-deficient soils or chronic anaemia in free range systems.

Another reason why the number of weaners reared per sow is lower for organic producers than conventional producers is that organic pig producers wean their piglets at eight weeks, compared with the three to four weeks in conventional systems. This delays remating and so lengthens the interval between farrowings. The organic sow must also have the ability to rear its offspring to eight weeks and maintain body condition.

Weaning age

By weaning at eight weeks old, compared with the extreme of two weeks old, which is done conventionally, the piglet is subjected to less of a sudden dietary change, because by this age more than 50% of its diet is likely to be creep feed. Both the sow and the piglets are also less liable to be stressed by the separation.

Growth

The objective is to achieve steady, uninterrupted growth with no check following weaning and with liveweight gain accelerating steadily over the period (Fig. 4.2). Mortality from birth should not exceed 5%, and by the end of this phase the young pig should have a strong, well-developed skeleton, a high proportion of lean meat, and minimal fat. Growth rate per day accelerates from 214 g in week 1 to 640 g in week 9.

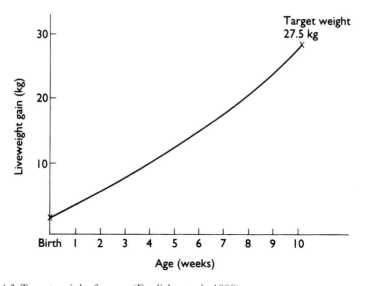

Fig. 4.2 Target weight for age (English, *et al.*, 1988).

Pigs have a deep body temperature of 39°C and will go to great lengths to maintain this temperature. The more energy needed for maintaining normal body temperature, the less there will be for growth in cold conditions, and in hot conditions pigs will reduce their food intake. This is why outdoor pigs grow less quickly than those in a carefully controlled indoor environment, but this may also be a reason why flavour is better.

As bodyweight increases from 10 kg to 100 kg, the daily intake of feed increases from 10 MJ (megajoules) to 50 MJ (Fig. 4.3). Daily gain increases from 600 g to 700 g up to 45 kg. From this weight boars and gilts gain at different rates. The organic pig producer, like most pig producers, does not castrate male pigs. Gilts continue to grow at the rate of about 700 g per day from 45 kg to 105 kg, whereas boars will grow at the rate of 800 g per day from 45 kg to 105 kg liveweight (Fig. 4.4).

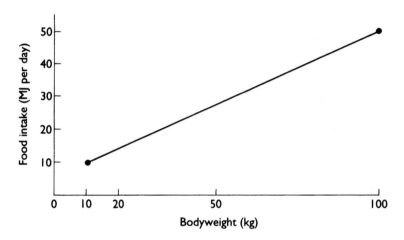

Fig. 4.3 Relationship between bodyweight and food intake.

Hansen, *et al.* (2000) described a Danish experiment in which organic pigs, kept outdoors, were either fed 70% organic concentrate plus organic clover/grass silage *ad lib*, or organic barley/pea silage. These were compared to groups fed 100% organic concentrate with access to an outdoor area, or 100% conventional concentrate with no access to outdoors. The low concentrate groups grew 14% to 16% slower than the 100% concentrate groups, but growth rate was still high.

Jensen, *et al.* (2002) experimented in Denmark with a tented system, plus access to grassland, and recorded growth rates of 780 g per day from 14 kg to 99 kg, with 8.9 weaned pigs per litter and a feed efficiency of 2.7 kg feed per kg of gain.

The perceived risk of taint with boars increases with age and weight. Because the organic producer markets only boars or gilts, the boars tend to be sent for slaughter at higher weights, that is up to 70 kg

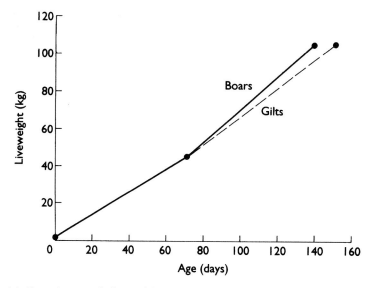

Fig. 4.4 Growth rate of gilts and boars.

deadweight. The liveweight and deadweight of pigs for different markets is shown in Table 4.1.

As the pig grows, it deposits progressively more fat relative to lean tissue. Thus food conversion efficiency decreases from 2.4 at 45–50 kg liveweight, to 2.76 at 68–73 kg, to 3.45 at 95–100 kg. Food conversion efficiency is the amount of food required per kg of liveweight gain.

Food conversion efficiency outdoors and the value of grass

Outdoor sows can be as productive as those kept indoors, but their nutritional requirements are different. The higher feed requirements of outdoor sows result from their higher maintenance energy

Table 4.1 Liveweight and deadweight of pigs for different markets.

Liveweight range (kg)	Deadweight range (kg)	Market
58–68	40–50	Light pork
69–95	51–70	Pork, cutters
80–102	59–77	Bacon
95–106	70–80	Heavy cutters
106 +	80 +	Heavy hogs

(English, *et al.*, 1988)

requirements, which are associated with a greater level of activity and the more variable environmental conditions to which they are exposed. Estimates of how far a sow walks per day range from 1 km to 10 km. It has been calculated that an additional 1 MJ of metabolisable energy per day would be used, approximately, for each km walked.

The sow's environment comprises a number of components, such as temperature, air movement, radiation, humidity, rainfall, snow, the absence or presence of bedding, shelter, shade and the opportunity to wallow. The additional energy requirements associated with the environment can be calculated. If the average critical temperature for an outdoor sow is assumed to be 15°C, then there are about 1600 degree days per year when the temperature falls below 15°C and, for a sow of bodyweight 160 kg, this means an additional energy requirement of 976 MJ ME (metabolisable energy) per year, or 17% higher than the annual requirement of a similar sow kept indoors.

Additionally, at very high temperatures, above 30°C, there was a 50% depression in feed intake compared with a controlled environment of 20°C. Although the piglets had access to creep feed, they did not compensate for a probable decrease in milk supply by eating more creep feed. Thus the weaning weight of the litter decreased by about 10 kg when animals were kept at 30°C rather than 18°C (Close, 1990).

It is also important to maintain sow weight during lactation so as to keep the breeding interval short. With gilts losing 23 kg during lactation, only 69% showed oestrus 21 days after farrowing, compared with 99% showing oestrus that only lost 2 kg during lactation, owing to a higher level of feed.

Animals kept outdoors have the opportunity to graze, but information about the feeding value of grass for pigs is difficult to obtain. Sows may also consume substantial quantities of bedding, especially straw, which will have some feeding value, but these extra bulky feeds should be treated as a bonus rather than as part of the feeding strategy. As mentioned later in Chapter 6, research is beginning to focus on the cheapening of nutrition for outdoor pigs.

When feeding outdoor gilts, the aim is to ensure sufficient fat reserves to compensate for adverse environmental conditions. Animals should therefore be grouped at an earlier age than with indoor gilts, and fed more. During oestrus, nutrition should be manipulated to obtain as high a litter size as possible. During pregnancy, feeding strategies should minimise loss of body tissue, and during lactation feeding should be *ad libitum*, to ensure minimum weight loss of the sow. The use of bedding, insulated huts, wallows, shade and appropriate nutritional management can minimise environmental effects.

At the moment, Soil Association regulations permit organic pig producers to feed 30% of the daily ration from non-organic sources. This means that the organic pig producer has to feed 70% of the daily ration as organic feed, and this is normally more expensive. The 30% 'non-organic' has to be from permitted sources.

Because concentrates form such a high proportion of the pig's diet and because organic concentrates are expensive, the supply of organic pigs is well below market requirements (Barrington, 1999).

In organic systems weaners are fattened outdoors. The groups are weaned at eight weeks of age, and the sexes are separated. Although fattening efficiency is reduced by 15% in the winter, growth rate is maintained at the expense of higher food intake. It has been observed that meat quality and texture are superior from pigs fattened outdoors on greenstuff, but this has not been the subject of a definitive experiment.

Outdoor housing

This has become increasingly popular, one reason being that the capital cost of setting up an outdoor breeding unit is about half that for indoor production, and the return on capital invested has been consistently better. Beynon (1990) suggests a stocking rate of 20 sows per ha (8 per acre) with nose ringing to avoid too much rooting about in the soil. The Duroc × Landrace is recommended as a hardy breed. Not all fields are satisfactory, the best being those with good, free-draining soils. Because drugs cannot be used, outdoor organic pigs have to be moved frequently, to prevent the build-up of parasites. It has been found (H. Browning, pers. comm.) that moving every three months is satisfactory. On the positive side, the pig enterprise becomes part of the whole farm rotation, and use is made of the nutrient fertility associated with outdoor livestock. This is in contrast with the indoor producer, who has severe problems of accumulated muck disposal. Outdoor pig keeping fits well into an arable rotation, with pigs put onto the last year of the ley so they can start ploughing up the grassland, although from a hygiene point of view it is wise to avoid returning to the same field within four to five years.

Carcass characteristics

Consumers of organic pig meat like small joints of pork, which means slaughtering porkers at about 70 kg liveweight to produce a 50 kg deadweight carcass. For organic bacon, they like a big eye and a big ham, which means a carcass weight in the range of 59–77 kg deadweight.

Hansen, *et al.* (2000) found that pork chops from organic pigs fed concentrates and roughage became less pale during storage than chops from conventional pigs, fed just concentrates.

The food conversion efficiency of the growing, finishing pig becomes less favourable the larger the finisher becomes. So the higher the slaughter weight the greater the feed cost per kg gain. However, the bigger the pig, the bigger the profit, unless the price per kg of meat is considerably higher for the lighter porker than for the bacon pig. And in this case the price differential is not usually sufficient to compensate for the lighter carcass. Furthermore, the labour costs involved in slaughtering and processing a light carcass are much the same as for a heavy one.

Health

Organic pigs do not routinely have their teeth cut, nor their tails docked. They are also generally healthier because they are kept outdoors, with much less build-up of infection because the pigs have more space and are moved to clean areas at frequent intervals. Because weaning is delayed until eight weeks of age, the weaners are more robust and are less likely to suffer from problems such as scour. Although the use of drugs such as antibiotics is banned, homoeopathic remedies have been found to be very successful.

Organic regulations

Those that have not been mentioned so far are:

- Housing – this must allow sows to express their full range of normal behaviour patterns, and must not involve permanent confinement or any housing system which prevents the sows getting up, lying down and turning round without difficulty.
- Stable social groups of gilts or sows with a maximum of ten animals per group.
- Fattening pigs on grassland where conditions permit, and retaining weaners in family groups.
- Where outdoor conditions do not permit, then with indoor management of fattening pigs housing must provide ample dry bedding with plentiful natural ventilation and light, outside dunging, rooting and exercise areas and a group size not exceeding 30.

The use of antibiotics, copper diet supplements and probiotics for growth promotion is prohibited.

Poultry production

The crux of the problem for organic poultry keepers, as with organic pigs, is that modern egg and broiler systems are extremely economical, with low costs of production. As production costs have been reduced, so welfare problems have been exacerbated, with layers being increasingly confined. Sainsbury (1992) estimated that the cost of free range egg production, with birds stocked at 1000 per ha was 50% higher than for the laying cage with 450 cm^2 per bird. The Soil Association regulations state that the stocking density of free range birds must not be greater than 250 birds per ha, which increases the costs still further. No wonder that, because organic producers insist on scrupulous attention to welfare, they need and ask for a higher price per egg and per fowl.

Changes

The egg laying performance and weight gain of table birds have both increased markedly, but so has the proportion of hens kept in battery systems. In 1960 the number of eggs per bird per year averaged over all systems was 185; in 1990 it was 253 per bird per year (Fig. 4.5). The average per free range bird was 220 whilst that of the battery hen was 290.

The assumption made in 1997 for egg production per bird per year was 270 (organic), 276 (free range), 282 (barn-reared) and 290 (battery) (Lampkin, 1997). The barn or straw yard system (Balty, 2000) is a semi-intensive system based on deep straw litter in the shed and the adjacent yard. There are communal nests on the floor and an enclosed pit for droppings. The larger scale organic egg producers used insulated barns holding up to 1000 birds, some with slatted floors and some with part slats and part litter. On a conventional farm it is unlikely that many of the buildings would be suitable for organic production, as the buildings are too big, being built for many thousands of birds.

In 1960, 31% of laying birds were kept in free range systems, 19% in battery systems and 50% in deep litter systems (Fig. 4.6). In 1990, battery systems had increased to 85%, free range had decreased to 13% and deep litter systems had almost disappeared, down to 2%

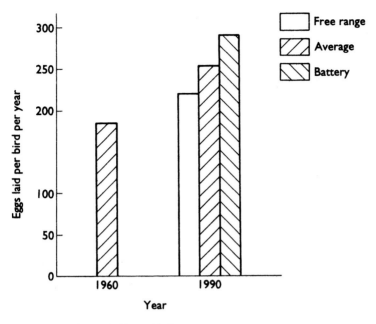

Fig. 4.5 Number of eggs laid per bird.

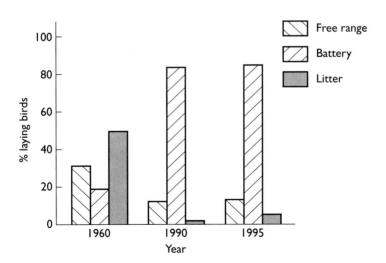

Fig. 4.6 Change in poultry-keeping systems for laying birds.

(Sainsbury, 1992). By 1995 proportions had changed very little, although there had been a slight increase in deep litter and barn systems to 4% (Lampkin, 1997).

In 1952 a table bird of 2.0 kg consumed 6.0 kg of food in 13 weeks; in 1992 a bird reached the same weight in only six weeks and ate a mere 3.8 kg of food. Organic poultry producers cannot hope to match these figures. In order to observe the much stricter welfare code promulgated by the Soil Association, they have to concentrate on a quality market and abjure mass production.

In order to ensure that the poultry in their charge have 'the opportunity to display normal patterns of behaviour', the costs of more space, more labour and more food must increase. An indication of the different costs involved in rearing organic poultry compared with conventional is shown in Table 4.2. The organic chick takes two weeks longer to reach 2 kg and eats more food, which is also more expensive per kg. In France the demand for Label Rouge poultry doubled during the 1990s. Because conventional, fast-growing poultry taste bland, with low fat content and reduced textural qualities, some of the criteria for Label Rouge production are 'slow growing strains, restriction of the use of fats and oils in the rations, a minimum of 75% cereals in the growing ration and a minimum slaughter age of 81 days' (Yeats, 2001). Housing and labour are more expensive because the organic birds have more room per bird and receive more attention. Finally, killing is done on site with the organic unit.

Desirable characteristics
The qualities of a good layer include a high yield of large eggs with both contents and shells of good quality, good food conversion for

Table 4.2 Production costs of poultry.

	Conventional (£ per bird)	Organic (£ per bird)
Cost of chick	0.24	0.24
Cost of feed (0–6 weeks) from birth to 2 kg		
(conventional 3.8 kg food at 15p per kg)	0.57	
(organic 5 kg food at 25p per kg)		1.25
Housing and labour costs on 100 bird unit	0.70	1.10
Killing, plucking and cleaning	1.00	1.50
Total	2.51	3.85

(Karl Barton, pers. comm.)

maintenance and production requirements, good liveability, resistance to disease, absence of broodiness, a docile temperament and good feathering. The stock from which replacements are bred must produce large numbers of hatchable eggs and vigorous chicks.

Size of the organic chick is increased by crossing a red hen with a large cockerel such as the Maran. This dual-purpose cross makes a good egg layer and a large table bird, which is important because, if you are breeding your own replacements, 50% of the chicks will be cockerels.

Table birds must have quick growth, good food conversion, good conformation, correct colour of flesh and feathering, good liveability and resistance to disease.

Size of unit

Work has shown that as unit size increases from as few as 20 birds to as many as 30 000, productivity falls linearly. Birds from groups of 20 reached 2.1 kg at the same age as those from groups of 30 000 reached 1.4 kg. The birds were similar genetically, ate the same food and there was no obvious sign of disease. The decline in weight was thought to be due to increased disease, but not to obvious clinical infection. It has been reported that some animals can produce substances that inhibit the growth of smaller animals of the same species. The result is that the more rapidly-growing animals survive at the expense of the slower-growing ones. This phenomenon could be a factor in the depression of overall weight in large groups, in addition to the problems of disease and behavioural stress.

As with other farm livestock, birds with the same apparent housing and nutrition facilities, and of the same genetic material, give vastly different results depending on the level of management. In surveys it has been clearly demonstrated that the farms which head the productivity league tables are the smaller units managed by the owner.

The Soil Association Standards recommend not more than 100 laying birds per housing group, and not more than 200 fattening birds per housing group, and that for laying birds artificial lighting must not prolong the day length beyond 16 hours. The amount of space per bird depends on its final weight. Thus if the final weight is to be 2 kg then eight birds per m^2 is sufficient, but, if the birds are to be taken to 3.6 kg final weight, then more space must be allowed for.

With a 100-bird unit it is important to ensure good ventilation, which keeps the birds cool and prevents the build-up of dust. This reduces the likelihood of disease. The shed should have two sides that open completely. Windows are not sufficient.

Beak clipping and all other mutilations are prohibited. Poultry owners, by beak clipping to prevent feather-pulling or cannibalism, are removing those tactile cells (taste buds) on the edge of the beak which allow the bird to decide whether to accept or reject food. To stop feather pecking in organic birds through boredom, it is sensible to keep them interested and active. This can be done by hanging a water-filled car tyre at poultry-head height, and putting cabbages or other greenery in the roof, which the birds can eat or play with.

Free range table birds

These are reared very differently from the intensive broiler bird. The free range bird is red feathered, slower growing and maturing, so that it takes nearly twice as long to reach a similar weight. The first two to three weeks are spent on straw or wood shavings and then the birds are given free access to pasture. It is essential to invest in either electric or chainlink fencing to prevent vermin, such as foxes, from killing the birds at pasture.

With the extra time taken to mature, the flavour is enhanced and the fat level is reduced. The bird has a more traditional chicken flavour but, as mentioned previously, will cost more.

Nutrition

Protein is required for general body development in all growing birds, and layers also need a good proportion because an egg contains 13–14% protein. Birds reared outside on pasture are usually fed concentrates and grain to provide about half the diet, the remainder being obtainable from good quality pasture. The grain, particularly barley, must be milled and not fed whole, because otherwise the eggs will have a slightly unpleasant taint.

Birds eat sufficient food to satisfy their energy intake, but this does not mean that they will eat enough protein, unless the protein proportion in the rotation is high. Protein quality is also important, the two most essential amino acids being lysine and methionine. The best quality protein for all poultry is white fishmeal (banned only for ruminants by the Soil Association). The best vegetable protein is soya bean which is low in methionine, but this can be made up by using sunflower meal in the ration.

In Japan, Takeuchi (2000) demonstrated that feeding an organic supplement to organic table birds reduced mortality from 7.8% to

1.4% compared to controls (with no supplement and no antibiotics), and reduced mortality from 2.8% to 1.4% compared to those receiving antibiotics. The final weight of those receiving the supplement was also higher (2.95 kg) compared to 2.88 kg for those receiving the antibiotics and 2.58 kg for those receiving neither supplement nor antibiotics.

Kohler, *et al.* (2000) in Germany compared the egg yolk luminosity and plumage condition of hens either reared indoors or at pasture. They found that intensive elements such as indoor cages and behavioural disorders led to low luminescence of the yolk, whereas fresh grass, exercise and space lead to a high luminescence of the yolk, which is correlated with intact plumage.

Measures of efficiency

These are equally applicable for all poultry producers, including organic poultry producers, though the financial targets will look slightly different. The efficiency measures are defined as follows:

- Liveability: the percentage of chicks alive at the end of the crop.
- Food conversion efficiency: divide the total weight of feed used by total liveweight at the start.
- Feed cost per kg liveweight: multiply food conversion efficiency by cost of feed per kg of liveweight.
- Average weight per bird: divide the weight of birds sold by the number of birds sold.
- Gross margin per unit of floor space: subtract the feed costs from the gross income and then divide the result by the floor area.

The Soil Association regulations give definite guidelines for space requirements for layers and fattening birds. Additionally, it is emphasised that all egg production systems must be planned to allow the birds to have continuous and easy daytime access to open-air runs, except in adverse weather conditions. The land to which the birds have access must be adequately covered with properly managed and suitable vegetation, i.e. grass of high digestibility.

Disease

The diseases to which poultry are subject are legion (Sainsbury, 1992). However, he does make the point that if vaccines and medicines were not so freely available, which they are not for organic producers, then health would be better because the poultry producer would be forced to rely on good housing, hygiene, husbandry and management.

It has been suggested that pullets reared outdoors make better use of pasture and may well make healthier adults. If the routine use of coccidiostats is to be avoided, pullets must be raised on clean pasture and great vigilance must be used to spot the first signs of disease. The Soil Association recommends that pastures should be rested for one year in three where set stocking is practised, unless stocking densities are low enough to prevent damage to the grassland and avoid disease build-up.

The European countries with the most organic poultry in 1996 (Foster & Lampkin, 1999) were Denmark (314 000), Austria (270 000) and Germany (262 000). Great Britain had 89 000.

Turkeys

The world production of turkeys is rising consistently. The smaller birds (4.5–6.5 kg) are kept rather like broilers, though not by organic producers, whilst the larger birds (8–15 kg) are reared in simple housing of the pole-barn type or even outdoors. Clean land which has not carried poultry or turkeys for two years is essential. Turkeys are extremely hardy and on range they need only simple housing or shelters.

When young turkeys are bought in they must be wormed at six and ten weeks of age to prevent the parasitic disease known as blackhead. It is important to give turkeys as much space as possible, and to partition the shed so that the turkeys which are eating do not disturb the ones that are resting. Ventilation must also be good. The straw in the pen must be kept very clean. If the turkeys get dirty then they start to grow replacement feathers, which are known as spars. These spars make plucking more difficult. After the turkeys have been finished, it is sound practice to keep cattle in the turkey shed. The cattle will convert the turkey straw into fertile compost. A dim night light is also important. Otherwise, when there is severe weather such as high wind, the turkeys will panic if they are in the dark, and may pile up, with some getting suffocated.

Ducks

The market for table ducks is steadily increasing, but egg layers are declining because of the strong flavour of the eggs. From four weeks onwards, ducks are reared outside, preferably on light sandy soils, as found in parts of Norfolk. Once again the outdoor system, particu-

larly in winter, is not as economic as the controlled indoor environment. In the successful management of ducks, good housing, coupled
with high standards of nutrition, are necessary factors for a profitable
programme, but hygiene is also important.

Geese

The basic nutritional requirement for the profitable farming of geese is
to provide quality grass which is short, leafy and digestible. A
reasonable maximum stocking rate for permanent pasture is 100 geese
per ha. However, a rotational system, in which the geese are moved
every four weeks to new ground, is likely to be better. Table geese will
be ready in 10–12 weeks, especially if good quality supplementary feed
is provided. During the winter months, the birds will need to be fed a
mixture of wheat, barley, maize and dried grass, plus adequate
minerals and vitamins.

References

Balty, J. (2000) *Practical Poultry Keeping*. Northbrook Publishing Ltd,
    Midhurst.
Barrington, L. (1999) Livestock. *Organic Farming*, **62**, 10.
Beynon, N. (1990) *Pigs – A Guide to Management*. Crowood Press, Swindon.
Close, W.H. (1990) *Outdoor Pigs*. Chalcombe Publications, Maidenhead.
English, P.R., Fowler, V.R., Baxter, S. & Smith, W. (1988). *The Growing and
    Finishing Pig*. Farming Press, Ipswich.
Foster, C. & Lampkin, N. (1999) *Organic Farming in Europe: Economics and
    Policy*, Vol. 3, Univ of Hohenheim.
Gruber, T., Tiefenbacher, R. & Baumgartner, J. (2000) Present status of pig
    fattening on selected organic farms in Austria. *Proceedings of the 13th
    International IFOAM Scientific Conference*, Basel, Switzerland, p. 365.
Hansen, L.L., Bejerholm, C., Claudi-Magnussen, C. & Andersen, H.J. (2000)
    Effects of organic feeding including roughage on pig performance, tech-
    nological meat quality and the eating quality of pork. *Proceedings of the
    13th International IFOAM Scientific Conference*, Basel, Switzerland, p.
    288.
Janmaat, L. & Groeniger, C.O. (2000) Organic pig husbandry in the Neth-
    erlands. *Proceedings of the 13th International IFOAM Scientific Con-
    ference*, Basel, Switzerland, p. 367.
Jensen, H.F., Andersen, B.H. & Hermansen, J.E. (2002) Concept for One-
    Unit Pen on Controlled Outdoor Areas Integrated in the Land Use.
    Proceedings of the 14th IFOAM Organic World Congress, Victoria BC,
    Canada, p. 86.

Klumpp, C. & Haring, A.M. (2002) Finishing Pigs: Conversion is more than respecting the Standards. *Proceedings of the 14th IFOAM Organic World Congress*, Victoria BC, Canada, p. 82.

Kohler, B., Folsch, D.W., Strube, J. & Lange, K. (2000) The influence of housing systems on the egg quality under particular consideration of the elements fresh grass and lighting conditions. *Proceedings of the 13th International IFOAM Scientific Conference*, Basel, Switzerland, pp. 289–292.

Lampkin, N. (1997) *Organic Poultry Production*. Welsh Institute of Rural Studies, University of Wales, Aberystwyth.

Sainsbury, D.W. (1992) *Poultry Health and Management*, 3rd edn. Blackwell Science, Oxford.

Takeuchi, M. (2000) The mortality of organic chicken. *Proceedings of the 13th International IFOAM Scientific Conference*, Basel, Switzerland, p. 372.

Yeats, W. (2001) Slow strain coming. *Organic Farming*, **71**, 14–15.

5   Arable Production Systems

Crop plants provide 70% of the food for the world's population and in addition produce vegetable oils and valuable raw materials for industrial processes.

In the Soil Association regulations it is stated that the basic characteristic of organic systems is the enhancement of biological cycles, involving micro-organisms, soil fauna and plants. The Soil Association encourages sustainable crop rotations, the extensive and rational use of manure and vegetable wastes and the use of appropriate cultivation techniques. Fertilisers, in the form of soluble mineral salts, cannot be used, nor can agro-chemical pesticides, chemical and hormone herbicides, nicotine, seed dressings based on mercurial and organo-chlorine compounds, aluminium-based slug killers or any synthetic pesticides. Also banned are continuous cereal rotations and the use of animal residues and manures from non-ethical livestock systems.

Rotations

A sound crop rotational system is of basic importance to the organic farmer. It reduces the risk of diseases and pests associated with monoculture, gives better control of weeds, spreads the labour requirements more evenly over the year, reduces the financial risk if one crop yields or sells badly and provides more interest for the farmer. For instance, take-all is a disease that cannot even be controlled chemically. Winter crops are more susceptible to take-all than spring sown crops and wheat is more susceptible than barley, so the safest position for winter wheat is immediately after a break crop.

Some crops can be chosen in organic rotations that control weeds through shading, such as winter wheat, if attention is paid to row width and drilling direction (Drews, *et al.*, 2002). Koocheki, *et al.* (2002) in Iran compared an organic cropping system with a high input one and found more weed species in the organic system, but weed biomass showed the reverse trend. The number of weed seeds was also higher in the organic system, which was attributed to the use of animal manure.

Winter barley is an ideal entry for oilseed rape because the earlier harvest allows early drilling of rape. It has been shown on organic farms that building up fertility by using a green manure crop, such as red clover, has paid off financially, with the subsequent boost in yield of the following cereal crop.

Varieties

Varieties are very different in their susceptibility to disease, and this is of great importance to the organic farmer. It makes sense to consider a varietal diversification scheme and to select at least three varieties for every farm, so that each has a different resistance factor. These different varieties can be sown in adjacent fields.

Weeds

The organic farmer does not want to be distinguished from his neighbours by having the weediest field, just because he cannot use chemical sprays. It is not just a question of aesthetics, but one of lowered yields and greater nuisance. Weeds compete with crops for nutrients and water, they shade and smother crops and can spoil the quality of a crop and so lower its value; and they can act as host plants for various pests and diseases. Weeds can interfere with cereal harvesting and some such as hemlock and ragwort are poisonous to stock.

In organic systems the number of hours of hand weeding needs to be reduced, as labour, particularly in the developed countries, is expensive. This can be achieved by competition, the timing of cultivations, adaptation of the rotation nutrient management system and by crop density (Kropff, *et al.*, 2000). Barberi, *et al.* (2002) experimented with a system of soil disinfection using hot steam to sterilise weed seeds.

The use of field margins around organic crops to increase the number of beneficial arthropods, which would then prey on harmful plant pests, was investigated by Pfiffner and Luka (2000).

Crop area

In Europe in 1996, the largest area devoted to organic arable crops was in Denmark (91 000 ha), with Italy next (62 000 ha) and then France (34 000 ha). This compares with 4800 ha devoted to organic arable crops in Great Britain (Foster & Lampkin, 1999). However, by 2003 this had increased dramatically to 44 413 ha (Soil Association, 2003). Changes in crop areas are shown in Table 5.1. Of the total crop area of 4 515 000 ha in 1996, wheat occupied 44% of the area. The area of spring barley had decreased by nearly 50% from the 1988 figure.

Table 5.1 Areas of crops and grass in the UK (per thousand ha).

	1988	1996	% change
Wheat	1886	1976	+4.8
Winter barley	856	749	−12.5
Spring barley	1021	518	−49.3
Other cereals	133	112	−15.8
Potatoes	180	177	−1.7
Sugar beet	201	199	−1.0
Oilseed rape	347	356	+2.6
Other crops	687	428	−37.7
Grass	6774	6665	−1.6

(MAFF census, 1996)

Nutrient removal and supply

A ready reckoner for the amount of N, P and K removed by certain representative crops is shown in Table 5.2. The requirement for P and K may be expressed in terms of the element rather than the oxide (P_2O_5 or K_2O). P_2O_5 contains 0.43 units of P; K_2O contains 0.83 units of K. The depletion of N, P and K from the grain of wheat, barley and oats is pro rata for yield, but the nutrient composition of the straw is different, oat straw containing very much more potassium than wheat or barley straw. Potatoes and kale are very much more exhaustive of N and K than the cereal crops.

Table 5.2 Nutrients removed by crops (kg/ha).

	N	P_2O_5	K_2O
Wheat			
grain 5 t/ha	93	43	30
straw 5 t/ha	17	7	40
Total	110	50	70
Barley			
grain 4 t/ha	67	33	22
straw 3 t/ha	17	4	31
Total	84	37	53
Oats			
grain 4 t/ha	67	33	22
straw 5 t/ha	15	9	74
Total	82	42	96
Potatoes			
30 t/ha	101	45	179
dry haulm 2.5 t/ha	50	6	112
Total	151	51	291
Kale			
fresh crop 50 t/ha	224	67	202

(Lockhart & Wiseman, 1993)

Farmyard manure and composting

Value to the organic farmer
The organic farmer needs all the farmyard manure that his animals produce: it is a valuable commodity for maintaining the fertility of his land, it is free and he cannot use most of the faster acting artificial fertilisers.

Most farmyard manure is produced by overwintering beef animals indoors, and as most beef and sheep farms are situated on permanent pasture in steep or rocky areas, it is the application of farmyard manure to grassland that is most usual, particularly to fields that are cut for silage or hay.

Poultry manure is another valuable source of nutrients to the organic farmer, but it must come from an ethical system.

Physical constituents of farmyard manure
Farmyard manure (FYM) consists chiefly of:

(1) Material that has been used as litter, mainly straw, though sawdust or bracken can be used.
(2) The food that has passed through the animal in an undigested condition.
(3) Urine, which contains that part of the food which the animal has digested but did not retain.

If the farmer wishes to use up as much straw as possible this will increase the litter proportion in the FYM; it will also increase the bulk and reduce ammonia, and therefore nitrogen loss.

Nutrient value of FYM
The three most important constituents of FYM are nitrogen, phosphate and potash. Additionally, there are the bulky organic parts derived mainly from the straw and the part of the food that has resisted digestion. Granstedt (2002) showed that the majority of the N (about 90%) in farmyard manure served to maintain the soil humus store and the long term capacity to supply nutrients.

Of the food fed to an animal, 70% to 90% of the nitrogen and phosphate reappears in the dung and 90% to 99% of the potash. However, a large part of this excreted N, P and K is lost during the making and storage of FYM. Most of the excreted N and K is in the urine (80% of N, 85% of K), whereas the phosphate appears in the dung. Thus two out of three of the most valuable nutrients are in the urine.

On average, dry FYM contains about 2% N, 0.4% P and 1.7% K, but different batches may contain very different percentages of nutrients depending on origin and storage. Percentage N can vary by a factor of two, percentage P by up to four, and percentage K by a factor of three.

Chemical analyses can measure the total quantities of N, P and K in FYM but not their availability to the crop. At most, all of the N in FYM is combined with organic substances and is released only when they decay; in practice about a third of the nitrogen is released quite quickly, but much is very resistant and persists in the soil. Much of the phosphorus is also combined with the organic matter but approximately half is quickly available. Most of the potassium is soluble in water and is quickly available to the crop.

A frequently quoted figure for the nutrient value of FYM from cattle is that 25 t/ha of wet FYM will supply 40 kg N, 20 kg of P and 80 kg K/ha.

In Switzerland, it was found that over a period of 6 years, the capacity of the soil in the plots fertilised with organic cattle manure to supply N to plants was greatly increased compared to soils which had been fertilised with artificial fertilisers (Langmeier, *et al.*, 2000).

Cattle manure

The values quoted above are for FYM from overwintered cattle, but the method of housing influences nutrient loss. With cattle in covered yards sufficient bedding must be used to absorb the liquid manure and there is little or no drainage: the dung is kept well compressed by the animals and fermentation is reduced to a minimum. Under such conditions very little loss takes place.

With cattle kept in open yards it is impossible to provide enough straw to absorb both the liquid manure and the rain falling into the yard, with the result that if the N and K are to be saved then drainage and storage of the liquid has to be provided.

Poultry manure

The composition of poultry manure depends on the kind of bird from which it comes, fattening fowls producing a richer dung than young or laying birds. The manurial value per ton will be greatly influenced by the amount of water or litter present, as these dilute the nutrients. Fresh poultry manure is liable to loss of ammonia by fermentation and of soluble constituents by leaching, but even so poultry manure is likely to be richer in nitrogen and phosphorus than FYM and its equal in potassium.

The case for composting

Composted dung is richer in nutrients and more active, provided it has been properly made, than comparatively fresh material that has not decomposed. However, in ordinary practice, manure cannot be rotted or stored without serious loss. Although well-rotted manure may contain more plant food than an equal weight of fresh manure, a given quantity of fresh manure will contain more plant food than composted manure that has been stored.

Shepherd, *et al.* (2000) suggested that one of the advantages of composting is weed sterilisation, but the disadvantage is that it will also increase the loss of nitrogen as ammonia. They suggested that there is a strong link with the amount of straw and thus the carbon:nitrogen ratio, which should be greater than 30%. Unfortunately, there is often a shortage of straw in many organic systems.

Causes of loss
Even when stored under the best conditions, FYM loses valuable nutrients, and under bad storage conditions this loss will be great. The chief causes of loss are:

(1) The escape of liquid manure and urine which contains a high proportion of the soluble nitrogen and the potassium.
(2) The escape of ammonia as a result of fermentation, particularly when the manure heap is exposed to the air and allowed to become dry.

Prevention of loss
The aim should be to get the manure into the soil or grassland as soon as possible. Remember that every time the heap is turned, fermentation is encouraged by the exposure to the air and ammonia is lost.

If shelter can be provided, that is beneficial, but if it cannot, then finishing the heap with a sloping top will help to throw the rain off, instead of washing it in and draining nutrients out.

Treatment of dung in the field
To save moving in the spring or to empty the yard, it is often necessary to make dung heaps in the field. The risk of loss from these heaps is greater than when properly sheltered in the yard. To reduce this loss, the heaps should be made on firm, level ground and if possible given a roof of earth. The siting of these heaps is important because the liquid manure and seepage must not be allowed to drain into water courses.

The manurial effect of dung is most evident in the first year of application but can still be beneficial, particularly on light soils, for up to four years.

The use of slurry
Liquid manure and slurry are often used on one or two grass fields that are close to the farm buildings. If possible this should be applied on a damp day, rather than on a dry, windy day when more ammonia will be lost.

Slurry is rich in nitrogen and potash but low in lime and phosphate. The result of over-application is to make the grassland coarse. This can be counter-balanced by the application of lime and an allowable form of phosphate.

Crop yields

It has been said that climate determines the crops that can be grown, and weather determines the yields. Seasonal factors such as weather patterns affect crop yields. There can also be a wide variation in yields between fields on one farm and even between different areas within fields.

A comparison of yields for some crops is shown in Fig. 5.1. The figures for organic yields are taken from fewer farms than are the figures for conventional farms. The slight reduction in yield that is apparent can usually be made up financially by the large premiums that have been available for organic cereals and by the demand for organic straw for thatching. Organic straw is sturdier and less liable to lodge, because the cereal receives no added nitrogen.

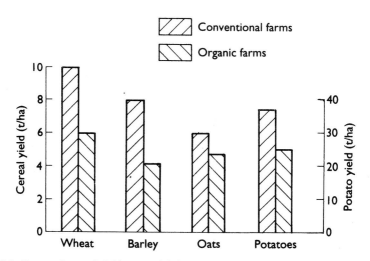

Fig. 5.1 Comparison of yield potential (Harper, 1983; Lampkin, 1990).

Arable crops

Wheat

Wheat is a deep-rooted plant which grows well on rich and heavy soils and in the sunnier eastern and southern parts of the country. Winter wheat can withstand most of the frosty conditions of this country, but is easily killed by water-logged soil conditions. It needs a pH higher than 5.5. It is the best cereal to grow when the soil is highly fertile, so it should follow grain legumes and root crops which have received manures and leave residual fertility. Most cereals are poor preceding crops for wheat, because they are host to the same diseases and pests.

It had been thought that older varieties of wheat, bred when fewer fertilisers and no agro-chemicals were used, would be more suitable for organic producers, but in fact modern varieties have been found to perform satisfactorily. However, trials are now underway to identify the main criteria for developing organic wheat strains (Angus, 2000). It is best to consult the National Institute of Agricultural Botany (NIAB) ranking list of recommended varieties annually, because varietal improvements are constantly being made. The use of a long-strawed variety to suppress weeds can be useful, but the most important factors in choosing a variety are yield and resistance to disease. Disease resistance can be enhanced by growing three different varieties in the same field, but it is essential that each variety is appropriate for the same purpose: for example, three breadmaking varieties, and to be able to prove this to the miller (M. Marriage, pers. comm.).

Although 90% of all commercial wheat being grown is winter wheat, the organic farmer should not be afraid of growing spring wheat, particularly if the winter sowing has been poor. Organic spring wheat intended for milling requires a protein content higher than 12%. In stockless ecological farming this can only be supplied by nitrogen from a previous legume crop. Wallenhammar, *et al.* (2000) in Sweden showed that if this legume crop was ploughed in early (July) then this resulted in a higher protein content and a higher yield of spring wheat.

Wheat grains can be processed to produce flour, which is then made into bread or biscuits. Bread-making and biscuit-making require different quality criteria. Wheat flour suitable for bread-making must contain adequate levels of protein of the right quality. Approximately 80% of the grain protein is in the endosperm, and this is easier to extract from hard rather than soft wheats. The protein forms a complex with water, known as gluten. Gluten has elastic properties and retains carbon dioxide gas bubbles during baking. This results in the popular open-textured loaf. Biscuits and cake are more readily made from a flour with a lower protein level.

Levels of the enzyme α-amylase in wheat grains also affect bread-making quality. Flour for bread-making requires low levels of α-amylase and this is favoured by a dry ripe dormant grain. The Falling Number (Hagberg) Test is used to determine α-amylase levels. A minimum falling number of 200 is normally used for defining flour for bread-making. The extra skill required in growing crops to meet the various quality standards is usually rewarded by a higher price. Mainly because organic wheat is the favoured arable crop of both bread and feed manufacturers, the area of organically produced wheat has increased from 6850 ha in 2001 to 14 394 in 2003 (Soil Association, 2003).

Barley

Barley is an important crop, with the best quality grains sold for malting and the remainder used for feeding all classes of stock, especially pigs, dairy cows and intensively fed beef. Barley straw can be used for bedding and as a maintenance ration. Barley is a shallow-rooted crop which grows well on chalk and limestone soils, with a preferred pH of 6.5. Its place in the rotation can be when soil fertility is low, which means that it can follow a previous cereal crop such as wheat.

Olesen, *et al.* (2000) compared the yield of spring barley grown with or without farmyard manure and with or without a catch crop. They found that grain yield in spring barley decreased in those plots without FYM or a catch crop. The highest and most stable yields were obtained where a mixture of ryegrass and clover was undersown in pea/barley and grown as a catch crop prior to the spring barley crop.

Spring barley usually yields better and has better quality grain than winter barley. The main advantage of winter barley is that it ripens very early, usually in July, and can be grown after grass. Barley has not been as popular as wheat or oats for organic growers, partly because winter barley has high nutrient requirements early in spring, before there is much biological activity in the soil, and partly because organic barley has not commanded the same high premium as wheat or oats. However, organic barley is a very useful crop for feeding to dairy and beef cattle.

A low nitrogen and high starch content is required in malting barley grains, and with the absence of nitrogen fertilisers the level of nitrogen in barley grown under organic conditions is likely to be low. Germination capacity, freedom from damaged grains and grain appearance can be controlled by careful choice of harvesting time and method and by the use of suitable drying temperatures. Before sowing barley or rye, it is important to discover what the market wants and to sign a forward contract with a miller. This avoids subsequent disappointment.

Oats

Oats are mainly used for feeding to cattle, sheep and horses (but not pigs, because of the high fibre content), but the best quality oats may be sold for making oatmeal, which is used for bread-making, oatcakes, porridge and breakfast foods, and this outlet has been very popular with organic producers and millers. Oats do best in the cooler and wetter northern and western parts of the UK. They will grow on most types of soil and can grow under moderately acid conditions (pH 5 or more). Oats can be grown at almost any stage in a rotation. They are deep rooted and can be productive where the soil is low in nutrients.

The different grain composition of wheat, barley and oats is shown in Table 5.3. Oats are higher in fibre and fat.

Table 5.3 Composition of cereal grains (g per kg dry matter).

	Wheat	Barley	Oats
Carbohydrate	786	781	698
Crude protein	124	108	52
Crude fibre	25	53	109
Fat	26	18	52
Mineral matter	18	31	29

Rye

Rye can be grazed early in the year and so is a useful crop for organic farmers who need early grazing for livestock and who cannot boost their grass with nitrogen in January or February. Rye will grow on poor, light acid soils and in dry districts where other cereals would fail. It is exceptionally frost hardy. Its place in the rotation is usually as a catch crop before kale or roots. Because of its extensive rooting system, autumn tillering and rapid spring development, it has a marked suppressant effect on weeds.

Triticale

This is a cross between wheat and rye. Triticale combines the yield and quality of wheat with the winter hardiness of rye and is also disease resistant. It can be used as a replacement for concentrates in a livestock ration because it is high in crude protein and essential amino acids. Its place in the rotation is similar to rye, so that it can be used as a forage crop or a green manure.

Mixed corn crops

The commonest type is a mixture of barley and oats. The yield of grain is usually better than if either crop was grown alone, but care must be taken to choose varieties which ripen at the same time.

Maize

Maize does best on fertile loam soils in the south-east of England. It is susceptible to frost damage and so needs to be planted late in the year, about mid-May. Its popularity is mainly because it is very high yielding and will make excellent high energy silage. Maize can make a useful break between cereals because it is not susceptible to the same diseases and pests.

Once the crop has become established it can be undersown with a legume, which will suppress weeds and benefit the soil.

Pulses

The grain or pulse forms of legumes have a high total protein content (20–26%) and can therefore be used as a natural supplement to cereals. Pulses are normally deficient in the essential amino acids methionine and cystine but contain enough lysine, whereas cereals are deficient in lysine but contain enough methionine and cystine.

Field beans

Field beans grow well on clay soils and heavy loams, provided they are well-drained and limed (pH above 6). Winter beans are not frost hardy and so are risky to grow, north of the Midlands. In some rotations, beans replace the clover break and they are usually followed by wheat. Yields for spring and winter field beans are shown in Table 5.4. The problem with field beans is their variable yield, which is partly caused by adverse weather conditions and susceptibility to pests and diseases (winter beans get chocolate spot, spring beans are aphid prone), and partly due to the uncertain activity of the necessary insect cross-pollinators.

Table 5.4 Yields of legumes (t/ha).

Winter beans	3
Spring beans	2.5
Marrowfat peas	3
Vining peas	4

Peas

Peas grow best on loam and lighter types of soil, provided they are well drained and limed (pH above 5.5). Most of the peas grown in this country are harvested dry for canning and packeting. Their yield stability is better than that of field beans. Growing peas and barley together makes a well-balanced silage, with the peas being high in protein and the barley supplying the energy. Threshing peas are often used as a cash crop break between two cereals.

Potatoes

The potato produces the heaviest weight of human food per ha, with the possible exception of sugar beet. In addition, diseased, damaged and small tubers, which are unsuitable for human consumption, are excellent as food for livestock. Each individual registered producer with the British Potato Council is allocated a quota and the same is true now for organic producers. National average yield for potatoes was quoted in 1981 (Spedding, 1983) as 37 t/ha, whereas the yield for organic producers was stated to be 25 t/ha (Lampkin, 1990) (Fig. 5.1). Early potatoes do best in early light soils in areas free from late frosts, whereas main crops grow best on deep loam soils. In a comparison between a biodynamic system, an organic system and a conventional one, the yield of potatoes in the biodynamic system was 62%, and the organic one 73% of the conventional one (Dubois, *et al.* 2000).

Potatoes can follow a grass ley, overwintered green manure, grain legumes or other crops which leave high levels of residual nutrients and organic matter. They are then usually followed by winter cereals which benefit from the residual nutrients from the applied farmyard manure. Pest and disease control are more of a problem with organically grown potatoes than are weeds. The main pests are wireworm and nematodes and blight, for which Bordeaux mixture can be used, but only if the level of copper in the topsoil is below 110 kg per ha.

Tamm (2000) pointed out that there is a ban pending on the use of copper, which will seriously affect production of organic potatoes.

Oilseed rape

Oilseed rape is very demanding on soil nutrients and no specialist organic market has yet been found for the organic product, so it is rarely grown in organic systems.

Linseed

Linseed has a lower nitrogen requirement than oilseed rape, so it could be grown satisfactorily in organic systems, but again no specialist market is available. It is also a crop that is difficult to keep weed free.

Sugar beet

The same applies to sugar beet as to linseed, and it is expensive to keep weeded.

Fodder crops

Fodder crops are useful in organic rotations; they are high yielding crops with high levels of metabolisable energy. They will also serve as a cleaning crop for weeds.

Mangolds

Ideally, mangolds require better soil conditions than swedes or turnips, and heavy loams will yield large crops (Table 5.5). The crop needs plenty of sunshine and hence is more popular in the south of England. The dry matter content is about 13% and the roots keep better than swedes; consequently they can be fed late in the winter. Mangolds are sensitive to acidity and yield best at soil pH levels of about 6.5. They are full-season crops grown in the root break in arable rotations, usually following and being followed by cereals. Due to their relative freedom from pests and diseases they can be grown frequently in rotations.

Table 5.5 Relative yields of forage crops.

	Dry matter yield (t/ha)	Crude protein (g/kg)
Mangolds (roots)	12.0	83
Fodder beet (roots)	13.5	79
Swedes (roots)	7.6	108
Turnips	3.3	192
Rape	4.2	200
Kale (autumn)	8.6	153

(Spedding, 1983)

Fodder beet
The advantage of growing fodder beet is that it produces a high level of dietary energy per ha (Table 5.5). Advances in breeding have produced over twenty bolt-resistant varieties with dry matter percentages varying from 13% to 22% (Goff, 1989). Fodder beet requires a pH of 6.5; because of the need to spread farmyard manure or slurry before ploughing in the spring, and the fact that harvesting will be in late autumn, land of light soil texture and strong structure with good drainage is essential. It is the risk of damage to soil structure by heavy machinery that is the major worry for organic producers.

Turnips and swedes
Turnips are generally lower in dry matter content than swedes and are divided into two main groups: white-fleshed varieties (8% dry matter), often called stubble turnips, which are frost susceptible and prone to bruising, but grow very fast and make valuable catch crops which can be grazed *in situ*; and yellow-fleshed varieties (8–10% dry matter) which are whole season crops and produce higher yields.

Stubble turnips were found to be almost the equal of white mustard as a catch crop and source of nitrogen to subsequent crops (Kotnik & Köpke, 2000).

Swedes are slower growing than turnips and are grown as full-season crops. The very high dry matter varieties may be so hard that they require chopping before feeding to animals to ensure high intakes.

Both swedes and turnips are sensitive to acidity and grow best at pH levels of 6.5. Like mangolds and fodder beet, the full-season turnips and swedes usually follow cereals in the rotation. Turnips can be grown from July to October, or they can be sown in July to provide autumn grazing before the frosts.

Kale, rape and cabbages
There are two types of kale: marrow-stemmed varieties which are high yielding, and thousand-head varieties which are leafier, more digestible and frost resistant. Bred hybrids between them combine many of the favourable characteristics of both. Kales generally out-yield rape, but the latter grow more rapidly and are more suited to catchcropping, like stubble turnips. Kale and rape can be grown in northern and upland areas but, because they are usually grazed, heavy soils in wet areas should be avoided. Utilisation can be

improved by strip grazing, especially when fed to cattle. Kale can be grown as a main crop, as part of a normal root break or as a catch crop following early harvested crops. Rape is usually grown as a catch crop, sown either in April for use in July, or sown in August for use after October.

Cabbages have a wide range of uses. They are a useful food for all classes of stock and, when suitable varieties are grown, high value crops can be grown for human consumption. These can be fed to stock when the market price is too low. They prefer moist, heavy soils, and seasons with plenty of rainfall.

References

Angus, W. (2000) Wheat breeding for the organic sector. *Organic Farming*, **66**, 7.

Barberi, P., Moonen, A.C., Peruzzi, A., Raffaelli, M. & Mazzoncini, M. (2002) Reducing weed seedling recruitment by soil steaming. *Proceedings of the 14th IFOAM Organic World Congress*, Victoria BC, Canada, p. 49.

Drews, S.J., Juroszek,, P., Newhoff, D. & Köpko, U. (2002) Competitiveness of Winter Wheat Stands against Weeds: Effects of Cultivar Choice, Row Width and Drilling Direction. *Proceedings of the 14th IFOAM Organic World Congress*, Victoria BC, Canada, p. 17.

Dubois, D., Mader, P., Gunst, L., Alfoldi, T. & Stauffer, W. (2000) Overview of the results of the third crop rotation period of the DOC-trial comparing organic with conventional arable farming systems. *Proceedings of the 13th International IFOAM Scientific Conference*, Basel, Switzerland, pp. 375–8.

Foster, C. and Lampkin, N. (1999) *Organic Farming in Europe: Economics and Policy*, Vol. 3. University of Hohenheim.

Goff, E. (1989) Fodder beet. *New Farmer and Grower*, **22** (15).

Granstedt, A. (2002) Use of Livestock Manure in Ecological Agriculture. Results from Field Experiments in Winter Wheat on Skilleby, Jarna, Sweden 1991–7. *Proceedings of the 14th IFOAM Organic World Congress*, Victoria BC, Canada, p. 11.

Harper, F. (1983) *The Principles of Arable Crop Production*. Granada, London.

Koocheki, A., Nassir, M., Zare, A. & Alimoradi, L. (2002) Weed Dynamics of Conventional and Ecological Cropping Systems in Different Rotations with Wheat. *Proceedings of the 14th IFOAM Organic World Congress*, Victoria BC, Canada, p. 59.

Kotnik, T. & Köpke, U. (2000) Efficient use of turnip (*Brassica rapa L.var.rapa*) for nitrogen management and enhancement of farm income. *Proceedings of the 13th International IFOAM Scientific Conference*, Basel, Switzerland, p. 81.

Kropff, M.J., Baumann, D.T. & Bastiaans, L. (2000) Dealing with weeds in organic agriculture – challenge and cutting-edge in weed management. *Proceedings of the 13th International IFOAM Scientific Conference*, Basel, Switzerland, pp. 175–7.

Lampkin, N. (1990) *Organic Farming*. Farming Press, Ipswich.

Langmeier, M., Oberson, A., Dubois, D., Mader, P. & Frossard, E. (2000) N fertiliser efficiency of cattle manure. Part 2: Influence of farming system. *Proceedings of the 13th International IFOAM Scientific Conference*, Basel, Switzerland, p. 82.

Lockhart, J.A.R. & Wiseman, A.J.L. (1993) *Crop Husbandry and Grassland*. Pergamon Press, Oxford.

MAFF census (1996), The Stationery Office, London.

Olesen, J.E., Rasmussen, I.A. & Askegaard, M. (2000) Crop rotations for grain production. *Proceedings of the 13th International IFOAM Scientific Conference*, Basel, Switzerland, p. 145.

Pfiffner, L. & Luka, H. (2000) Enhancing Beneficial Organisms with Field Margins – an Important Strategy for Indirect Pest Control on Organic Farms. *Proceedings of the 13th International IFOAM Scientific Conference*, Basel, Switzerland, p. 105.

Shepherd, M., Philipps, L. & Bhogal, A. (2000) Manure management on organic farms: to compost or not to compost? *Proceedings of the 13th International IFOAM Scientific Conference*, Basel, Switzerland, p. 50.

Soil Association (2003) *Organic food and farming report*. Soil Association, Bristol.

Spedding, C.R.W. (ed.) (1983) *Fream's Agriculture*. John Murray, London.

Tamm, L. (2000) The future challenges and prospects in organic crop protection. *Proceedings of the 13th International IFOAM Scientific Conference*, Basel, Switzerland, pp. 106–9.

Wallenhammar, A-C., Andersen, L.E. & Svaren, A. (2000) Organic production of quality spring wheat following a clover ley. *Proceedings of the 13th International IFOAM Scientific Conference*, Basel, Switzerland, p. 87.

6 Farm Size and Enterprise Combinations

An organic farm should be just as profitable an investment for the owners as a conventional farm. In the past, the image of organic farming has been of a small untidy farm that has been bought to provide a getaway from the pressures of modern life. This can be a valid aim, but more is needed to make an organic farm profitable than just the desire to escape to nature. Careful calculation should be made of the profitability of each projected enterprise, and this must include the cost of the two-year conversion process, during which time yields are likely to be below average and a great deal of learning will take place. Markets for organic produce also need to be sought, particularly when an organic premium is anticipated.

It may be argued that this is a counsel of perfection, that very few people going into farming have sufficient capital to ensure that the farm they buy or rent will be profitable, even if it has enough fertile acres, with perfect buildings and fencing. The point at issue is that before embarking on organic farming it is essential to calculate, as far as possible, that the farm has the potential to produce the required income. This at least would prevent a great deal of subsequent financial worry.

Farm size

Before going organic, the family must decide what standard of living they require from the farm and what income they are likely to require in the future. This is not to say that all the income must come from the farm; often there will be supplementary income from a different source, but the likely income from the organic farm must be calculated as accurately as possible. On large farms where several workers are employed it soon becomes clear whether, for instance, the sheep enterprise is paying the wages of the shepherd. It may be less obvious if the sheep are looked after part-time by a member of the farmer's family who, instead of receiving a fixed wage, takes a share of the profit. The other important point when considering farm size and enterprise combination is to take into account the different productivity from a hectare of fertile lowland compared with a hectare of steep Welsh mountainside. The difference may well be a factor of ten.

Some assets are difficult to cost accurately: for example, the notional financial value of providing a haven for wildlife, or of keeping your part of the river free from excess nitrates. Another example would be the value of keeping beef cattle indoors and then using the farmyard manure to improve soil structure and maintain wheat yield.

There is little point in defining standard of living here, because no two families have the same outlook or requirements, but the costing must be done in advance. Even the warmest glow of organic satisfaction is unsustainable when debt is mounting annually.

Enterprise combinations

An organic farmer needs to experiment with the appropriate mix of complementary cropping for his/her farm and to adopt a positive approach to animal and crop health. The best method of combating plant and animal disease is to prevent it occurring. Because the organic farmer cannot intervene, except under exceptional circumstances, with herbicides, biocides and anthelmintics, he/she must manage the farm so that the need never arises. Because the organic farmer does not have an easy, or permanent, solution he/she has to be that much cleverer and more watchful. Much more thought has to go into crop rotation and grazing management, which can only be beneficial. For a more detailed account of rotational strategies consult *Organic Farming* (Lampkin, 1990).

In planning crop rotations it is necessary to consider disease control, weed control and soil nutrient status, as well as finding out in advance which organic crops are likely to be the most profitable. For instance, organic oats have sold much better than organic barley; there is no market for organic sugar beet but there is for organic potatoes.

Early bite for sheep may be a problem, because no fast acting nitrogen can be used in February; however, the nutrition of the sheep flock could well be met by growing rye.

Some organic farmers find it difficult on heavy land to establish enough white clover in the grass break between cereals to fix adequate amounts of nitrogen. Red clover can therefore be considered because it establishes much more quickly. Unfortunately, it is also predisposed to stem eelworm and clover rot, so it is not advisable to grow red clover too frequently in the same field. It is often thought that because the organic farmer does not use fast-acting N, P and K fertilisers, then he/ she is oblivious to the nutrient status of the soil, but the reverse should be true. Farmyard manure and white clover must form the basis of soil nutrient control, coupled with a sound grazing management policy, which will ensure that nutrients are recycled. Table 6.1 shows that on a world basis the recycling of livestock wastes supplies more N, P and K (about six times as much K) than fast-acting fertilisers, although it is very difficult to avoid considerable losses of excreta nitrogen from volatilisation and leaching. However, there is still a world-wide need to make more effective use of nutrient in excreta.

Table 6.1 Inputs and outputs of nutrients for the world's crops 1979–80 (megatonnes).

	N	P	K
Inputs			
Natural fixation of N			
Arable cropping	50		
Grassland	100		
Fertilisers	57	14	19
Recycled livestock wastes	80	17	120
Output			
in arable crops	75	15	75
in grassland	200	25	200

(Cooke, 1987)

Animal systems

Although stockless systems can be practised on arable farms, by the use of green manures to maintain soil nutrient status, the need for the addition of animals as a source of recycled excreta, and as graziers, has long been recognised. At the time of the 'golden hoof', field owners paid shepherds for the use of their sheep, if only to have them penned up on a particular field overnight.

By housing cattle overwinter and composting the farmyard manure, the organic farmer has ready access to a balanced fertiliser that can be spread where most required. The grazing animal does not actually import fertility onto the farm but it does recycle nutrients where it grazes and provides a source of manure when housed. This is as true of sheep, pigs and poultry as of cattle. The only problem with outdoor pigs is that they tend to rip up pastures.

Dairy cows and sheep
On heavy wet land, grazing by weighty dairy and beef cattle may perhaps be limited to only six months of the year, because of poaching, whereas the herbage may well grow for nine months. This wet land is also unsuitable for heavy machinery. A sensible and profitable use of this excess grass is to graze it with sheep, particularly in the autumn and winter months. If the sheep belong to the same farm, then the use of cow pastures helps to defray winter feed bills and, if sheep have to be brought in from elsewhere, a rent can be charged.

There is the further benefit to the pasture that by grazing these swards with sheep during the winter the sward will be improved. The rather open pasture typical of dairy cows will be thickened up with an increased number of grass tillers. At the same time, by grazing the sward down to 3 cm height in the autumn the white clover will be encouraged to increase its number of stolons, because of exposure to the light. This will provide a base for expansion in the spring. The final benefit is that the grass will contain a higher proportion of young leaves by the spring, which will increase its digestibility.

Dairy farmers are often reluctant to allow sheep to stay on the dairy cow pastures into January, because they feel that this will reduce the amount of herbage available for the turn-out of the dairy cows in the spring, and will also reduce the amount of silage conserved in the important first cut. Sheep can certainly be left on dairy cow pastures into the New Year without reducing the amount of digestible herbage available for the cows, providing that they are not allowed to graze the sward below 3 cm and that they are removed by mid-February.

Parasite control in sheep
The control of internal parasites in sheep is acknowledged to be one of the harder problems in the management of both organic and conventional enterprises. The organic farmer is allowed to dose any animal that is seen to be suffering from parasitic infection, but only those animals that are suffering, not the whole flock. The animals that

have been dosed with anthelmintic cannot be sold as organic until twice the recommended interval. And even then many organic farmers feel that once an organic lamb has been drenched with anthelmintic it should never be sold as organic. No homoeopathic remedy has yet been shown to be effective against a heavy infection of internal parasites.

Much the preferred method of control is to practise clean grazing and so avoid the lambs ingesting too large a worm population before they have built up their own immune system. To do this effectively, the land needs to be rested from sheep for a whole year. The most commonly adopted system is a three year rotation on the grassland, with one year cattle, one year conservation and one year sheep. If the farm land is fertile enough to grow arable crops then this break will get rid of the parasites, and the sheep, particularly the lambs, can then graze the clean grass in the first year of the ley. With sheep and beef alternating, the sheep will also keep the pasture clean of parasitic larvae for the beef animals, with the exception of *Nematodirus* which is infectious to both sheep and young cattle. Clean grazing is also the best method of control for the conventional farmer, because some species of parasite are now building up resistance to several of the current range of anthelmintics.

Sheep and poultry
Some organic farmers who keep free range poultry and sheep have observed that they have had no parasitic problem with the lambs under these circumstances. The hypothesis is that the poultry destroy the infective larvae on the pasture by eating them. However, a recent experiment (Newton, *et al.*, in press) has shown that, so long as the land is rested from sheep for 8 weeks or longer, the number of free-living infective L3 larvae falls dramatically, whether the land is grazed by poultry or left ungrazed.

Enterprises and finance

Another factor in the successful choice of enterprise combinations is the blending of systems so as to produce a regular income. One of the reasons for the popularity of dairying has always been that it commands a monthly milk cheque. The return from lambs tends to be concentrated over a three-month period, whereas that from beef is slow and even more irregular. However, where the product is sold directly from the farm, it may well be better policy, financially, to sell the lambs over a longer period of time, six months instead of three. This makes management more difficult, but evens out the income from the sheep enterprise.

Finally, there is a consideration of the various grants available. Headage payments for sheep and cattle are available to all farmers, and there are grants for extensive farming and the reduction of stocking rates in certain areas. Some crops are also subsidised, irrespective of yield. There is little point in detailing the grants here, because they keep changing. For instance, although a grant is currently available from the government for converting to organic farming, the terms of this are likely to change.

References

Cooke, G.W. (1987) *Fertilizing for Maximum Yield*, 3rd edn. Blackwell Science Ltd, Oxford.

Lampkin, N. (1990) *Organic Farming*. Farming Press, Ipswich.

Newton, J.E., Adams, C.C., Bairden, K., Coop, R.L. & Jackson, F. (in press) The effect of grazing by poultry on the number of sheep parasitic larvae found on organic grassland. *The Veterinary Record.*

7 Organic Standards: Problems and Solutions

There is now a legal definition of 'organic' and a set of standards that has been agreed by UKROFS (the United Kingdom Register of Organic Food Standards), an organisation set up by the government to safeguard the public from the misleading labelling of food products. Organic food must not only be produced according to a single set of regulations, but the farms and smallholdings are inspected on an annual basis to ensure that the regulations are being adhered to. The organic symbol holder has to pay for the symbol and for the annual inspection.

It would be idle to pretend that certain of the regulations do not make farming harder or more expensive than for the conventional farmer, or that no organic farmer questions the sense of one or two of the regulations. After all, the regulations have been altered quite substantially and are still being updated, this time by the European Union (*Organic Farming*, 2000). There is no infallible logic behind the rules – mostly they arise from committee decisions, which tend to be compromises. However, the successful organic farmer succeeds mainly through a strong belief in the rightness of organic farming, as well as through hard work and skill, and also through perseverance verging on obstinacy. The sections below outline some of the problems, with perceived solutions where such exist.

Beef calves

Before the Soil Association Standards were changed in 1992 it was allowable to buy beef calves from non-organic herds (provided they were bought from a farm and not from a market and were less than 28 days old), put them onto an organic suckler cow and then sell them as organic when they had been suckled and reared on the organic farm for at least six months. This was very convenient for many organic farmers, because the calves could be bought locally.

Now the Standards have been tightened up and beef has been treated in the same way as sheep. This means that organic beef calves have either to be born organic on the farm or purchased from another organic farmer. Because the number of organic farms is still small, this makes the buying-in of organic calves more difficult, but the efficient and well-organised organic farmer can overcome this obstacle by entering into a regular agreement with other organic farmers. For instance, an organic farmer with a beef suckler herd can purchase the unwanted bull calves from an organic dairy farmer to replace calves lost to his own sucklers.

Beef

Because the organic beef farmer must feed 60% of the dry matter in all ruminant diets as either fresh green food or unmilled forage, then his animals are likely to grow more slowly than barley bull beef, for instance. The slower growth produces a darker red in the beef after slaughter and some supermarket managers have objected to this, claiming that the consumers associate this with tougher steaks. However, in a consumer trial (Lowman, 1989) to compare the appearance, cooking quality and eating quality of steaks from the two groups of animals, it was apparent that consumers preferred organic steaks in terms of overall eating quality. This suggests that the public needs more information about organic beef.

Pigs

Organic market share

The costs of running an organic pig farming unit are much higher than those associated with a conventional system. A survey from Scandinavia (Smith, 2001) has shown that the organic market is finite and that pricing organic pigmeat at a premium of 30% would result in a 4% to 7% increase in market share, whereas if the premium were restricted to 20%, then growth would be 19%. However, the number of organic pigs in Britain is still small, with only just over 1000 sows being registered organic in 1999.

Feed costs for organic pigs

The assumption is normally made that the feed value of grass to the outdoor pig is equalised by the extra energy expended in walking further outdoors and adjusting to a much higher range of temperatures. In a trial at the Dorset College of Agriculture (Machin, 1990), sows were rotated around grassland, with the animals being moved before they started to poach or overgraze the pasture. An electronic sow feeder with balancer feed (1 kg per sow per day) was available to the sows outside. These sows gained weight, produced 10.5 live piglets and were back in pig in 13.6 days. Further work was carried out offering either fodder beet *in situ* or maize silage, both cheap feeds, with satisfactory results. Other benefits from feeding low-nutrient, dense fibrous feeds have been:

- *Behaviour and welfare:* Stereotyped behaviour has been reduced, lying time increased, with the sows being slightly more content.
- *Increased fat content in milk:* The sow has responded to increased fibre in the diet, like the dairy cow, by increasing the production of acetate, which has then beneficially increased the fat content of the milk.
- *Increased litter size and weight:* Bulky diets have appeared to increase litter size and the birthweight of piglets.
- *Decreased lower critical temperature:* Because of the heat of fermentation, the sow appears to have a lower critical temperature when consuming fibrous feeds.
- *Improved health:* There have been suggestions of improved health, particularly in the digestive tract, through using diets containing bulky feed materials. For example, ulceration of the stomach may be reduced due to the feed having larger particles.
- *Increased appetite in lactation:* One of the most important benefits of bulky diets is that, by having the stomach stretched during gestation, sows appear to have an increased appetite during lactation, thus reducing weight loss.

Clearly much more research needs to be carried out on organic pig production to reduce feed costs.

Insulated huts
One problem that has been highlighted with the keeping of sows outside at farrowing has been increased piglet mortality caused by low temperatures. Research into the beneficial effect of insulating outdoor pig huts (Corning, 1990) indicated that temperature variation had been much reduced. A reduction of 1% in piglet mortality would justify an extra £100 on the cost of every farrowing hut, for a 200 sow unit.

Breeding for the ideal organic pig

Unfortunately, breeding for many of the ideal characteristics of the outdoor pig involves a trade-off with less desirable characteristics (Bichard, 1990). In the past, the outdoor herd owner was prepared to accept a relatively poor carcass with too thick back fat, in return for the animal's ability to survive without too much management, and having animals with high feed intakes. Coloured skin, which protects the sow against sunburn, and docile intelligent sows (also associated with coloured pigs) are associated with fatter carcasses. However, with the recent increase in the number of sows kept outdoors, it is anticipated that more genetic resources will be applied to solving some of these conflicts.

Poultry

Although indoor systems of egg production are much cheaper to run than outdoor organic systems, due to higher egg production per hen per year and to lower feed costs, poultry longevity can be increased outdoors, which will lower the cost of replacement birds. Thompson (1978) tested his future breeding stock in an outdoor environment which had carried hens continuously for many years, with a consequent build-up of disease. Those pullets which survived became the foundation stock.

The inference for organic poultry is the same as for organic pigs. The aim of the breeding policy must be to produce strains of hen that do best outdoors, where they are to live.

Dairy cows

Summer mastitis in dairy cows

In the Soil Association Standards it states that 'the practices employed in the management of livestock must be directed towards maintaining animals in good health and preventing conditions where the use of conventional medication becomes necessary'. A good example of this is the avoidance of summer mastitis in dairy cows by calving the herd in the spring. This has the added bonus of reducing the amount of concentrates required in peak lactation, because of the availability of highly digestible spring grass and also because of qualifying for the high summer milk price.

Sheep

Vaccination against clostridial diseases in sheep, such as braxy, blackleg and lamb dysentery, is restricted to cases where there is a known disease risk on the farm or neighbouring land, which cannot be controlled by any other means. It is recommended that if there is a known disease risk of lamb dysentery, for instance, then instead of using the seven-in-one vaccine, a two-in-one vaccine may be used. Unfortunately, there are no alternative homoeopathic treatments for the clostridial diseases. Many sheep farmers have been apprehensive in the first year after stopping using the routine seven-in-one vaccine. They expected catastrophic losses but this has not happened, mainly because the animals have been kept in good health by attention to nutrition and welfare.

The regular supply of organic lambs
In 1990 some supermarkets began to stock organic lamb on their shelves. Their stated requirement was for a regular supply of the same number of quality carcasses every week of the year. This quickly highlighted the problem of being only a small movement. At the time there were about a hundred organic sheep farmers with the Soil Association symbol. This was not enough to meet this regular demand for organic lamb, particularly in the difficult months of April, May and June. In order to get a regular supply, there has to be over-production and co-ordination of lambing times between the organic producers, and this degree of organisation simply does not exist. However, it is a problem that could be overcome if it were more widely accepted that the supermarket was potentially the major retail outlet for organic lamb, rather than the local butchers or at the farm gate. In turn, the supermarket must guarantee its contract with the farmers and not suddenly leave them in the lurch. If a larger proportion of the population is to eat organic produce, then it must be via the supermarket.

The absence of fast-acting nitrogen fertilisers
The answer to the absence of fast-acting nitrogen fertilisers, with arable crops and grassland, has been threefold: first, to make much greater use of farmyard manure, and to make sure that it is properly stored and not allowed either to be washed down the drains or to volatilise into the air; second, to make much greater use of the legume for its power of nitrogen fixation, particularly white and red clover

and crops such as field beans; third, to grow special crops, such as rye, that can be grazed early in the season, before grass/clover swards have begun to produce much growth.

The standard, conventional method of increasing the content of white clover in existing grassland has been to slot seed with white clover, following the spraying of a weak solution of paraquat onto the grass. The paraquat is used to check the grass and thus allow the young clover seedlings to develop without too much competition. Organic farmers, who cannot use herbicides such as paraquat, have found that by grazing the grass down very tightly or by setting the mowing blades very low when cutting for silage, slot seeding has been most effective. Survey data (Newton, 1992) has shown that the levels of white clover in permanent grassland and temporary leys have averaged between 20% and 30% over the whole growing season. The success of organic farmers in increasing the levels of white clover in their pastures has been considerably aided by their not applying regular doses of fast-acting nitrogen fertilisers.

Weed control

Because no herbicides can be applied to organic crops it is assumed that there will be very serious weed problems. Organic farmers rely on good husbandry, including rotations and delayed drilling when necessary. Weeds can be a most important cause of yield loss. The rotations employed, together with pre-sowing and within-crop cultivations, and the use of high seed rates, help to keep down annual weeds. With cereals, most growers find that the use of finger weeders, two or three times in the spring, can be quite effective in keeping down annual weeds, particularly in spring-sown crops. It is perennial weeds, such as thistles, that are the major problem. These can be controlled only through continually weakening them by cutting at the right time.

Disease and pest control

The most serious disease problems encountered in organic crops are blight in potatoes and seedborne diseases in cereals. Plant breeders are improving blight resistance in potatoes. As organically grown cereal seed cannot be treated, it is very important that it is tested for seedborne diseases such as fusarium and smut. Foliar diseases in organic cereals have rarely been a serious problem, when compared with conventionally grown crops (Lockhart & Wiseman, 1993).

Good husbandry can help limit pest problems. Frit fly damage can be reduced by allowing at least a six-week interval before drilling cereal seeds following a ley. Delaying drilling of winter cereals will also reduce the likelihood of aphid-transmitted Barley Yellow Dwarf Virus. Biological control methods are also being used. Many farmers are encouraging pest predators by habitat management.

References

Bichard, M. (1990) *Breeding for Outdoor Pig Production. Outdoor Pigs.* Chalcombe Publications, Maidenhead.

Corning, S. (1990) *Outdoor Pig Production in the UK. Outdoor Pigs.* Chalcombe Publications, Maidenhead.

Lockhart, J.A.R. & Wiseman, A.J.L. (1993) *Crop Husbandry and Grassland.* Pergamon Press, Oxford.

Lowman, B.G. (1989) *Organic beef production. Organic Meat Production in the 90s.* Chalcombe Publications, Maidenhead.

Machin, D.H. (1990) *Alternative Feeds for Outdoor Pigs. Outdoor Pigs.* Chalcombe Publications, Maidenhead.

Newton, J.E. (1992) *Herbage production from organic farms.* Research Note 8, Elm Farm Research Centre.

Organic Farming (2000) EU Livestock Regulations. *Organic Farming*, **66**, 8.

Smith, P. (2001) *Practical Pig-keeping.* Crowood Press, Marlborough.

Thompson, M.A. (1978) *The Organic Poultryman.* Matthew A. Thompson, Bridport.

8 Financial Management of the Farm

This chapter is for those people who are new to organic farming and those existing organic farmers who want greater control of their finances and more information for decision-making. Even a profitable farm can experience short-term cash flow difficulties, which could lead to business failure. This is why more attention should be paid to depth of resources, initially, and to regular monitoring of financial status. The farmer will also need to know how to determine whether a new venture is worth pursuing. Any new venture will require additional time, money and other resources but it is important to be able to get an idea of the profit potential for the new activity. Following the straightforward exercises outlined below should provide warning signals and help to prevent the nasty shock of suddenly being made aware of insolvency. The discipline of monthly accounting will also help farmers choose and change to the right financial option.

In managing a farm business two exercises are recommended. One is the monthly monitoring of cash flow, to determine how much money is coming in and how much is going out; and the second is the exercise of costing, which can be done over a set time period, for example annually or every six months. The costing exercise provides the farmer with recent information on the cost of producing the output over the given interval, which can then help him make decisions on issues such as pricing, future output and volume of output. Ultimately, costing helps to reveal not only how much overall profit or loss the farm is making, but also which particular enterprises or products are financially successful and which are not worth continuing, and to determine the profit potential of any recent venture. It is important that these exercises are accompanied by an awareness of how the different farm enterprises interrelate, so that the overall configuration is in balance.

Cash flow

At the beginning of each year, the farmer should draw up a chart (Fig. 8.1), which will enable him to see how much cash he has available each month. There are two main sections – income and expenditure. The expenditure section is subdivided into two sub-sections: operational

	Jan	Feb	Mar	Apr	May	Jun	Jul	Aug	Sep	Oct	Nov	Dec

Income

Sale of:
  1 Wheat
  2 Milk
  3 Straw
  4 Grants

  5 Total income

Expenditure

Operational:
  6 Rent
  7 Wages
  8 Seeds
  9 Machinery
  10 Vet and med
  11 Fencing
  12 Electricity
  13 Tractor fuel
  14 Telephone
  15 Advertising
  16 Stationery
  17 Insurance
  18 Fees
  19 VAT
  20 Repairs

Financial:
  21 Interest
  22 Loan repayments
  23 Leases

  24 Total expenditure

  25 Net cash flow (5–24)

  26 Opening balance

  27 Closing balance (26+25)

Fig. 8.1 Forecasting monthly cash flow.

expenditure, which relates to the farm itself; and financial expenditure, which depends on the level of borrowing. Weak operational cash flow may cause borrowing and interest payments to rise, so it is important to be aware of the link between the two sub-sections. The list in Fig. 8.1 is not meant to be exhaustive, but simply to illustrate the kinds of items that should be taken into account.

Any forecast of income requires a prediction of future output volumes by product category and their prices. With the withdrawal of so many guaranteed prices with the free market economy, the forecasting of prices is particularly difficult. It may be relatively easy for milk, and possibly wheat, but is much harder for volatile enterprises such as beef, pigs (which have never had a guaranteed price) and turkeys.

Different farm enterprises vary as to the regularity of sales. For instance, one of the attractions of dairying is the monthly milk cheque. This may vary in size depending on the stage of lactation in the herd, but it is rare to have a month with no milk to sell. With beef, on the other hand, particularly with suckler beef, there can be two years or more between putting the cow in calf and selling the finished animal. Income from beef is unpredictable, but production is far less labour intensive than dairying – the cows have to be milked twice a day, every day of the year.

Forecasting income and expenditure is not a popular exercise, partly because it may reveal a precarious future and partly because forecasting accurately is known to be difficult. To engage in an exercise that experience has taught one is unreliable, may seem a waste of valuable time. However, the exercise is crucial, and the uncertainty of price forecasting and predicting the volume and timing of sales can be mediated by carrying out three related exercises:

(1)   Take an optimistic view of the year ahead.
(2)   Take a pessimistic view.
(3)   Take a realistic view.

The first two steps will help to define the best and worst case possibilities, allowing the farmer to make a judgement on what is realistic. The realistic forecast, which may be an average or central case derived from the other two, is the one that should be proceeded with.

Check on monthly cash flow

Throughout the year a monthly check of the actual cash flow against that forecast at the beginning of the year is needed. If a negative cash balance is forecast for a month or more, or if the monthly cash flow is persistently negative, causing a serious depletion of cash reserves, the sensible strategy is to:

(1) Go to the bank in advance, to give warning of the problem. Informing the bank manager of the expected cash difficulties, before they occur, will show that the farmer is in control of the situation, and may improve his bargaining position. The cash flow forecast can be used to demonstrate that the problem is temporary, and that any new funding requirements can be covered. Larger businesses have most influence as will those farms with considerable capital assets, allowing them to take out secured loans more easily.

Overdraft facilities are a relatively expensive way of financing your enterprises. If your bank balance is consistently negative, it suggests that you are not generating sufficient cash to run the business, and that more permanent long-term funding is needed.

(2) Ask suppliers for more time to pay.

(3) Encourage buyers to pay sooner.

(4) Move unsold stock.

(5) Endeavour to postpone capital purchases such as new machinery until later.

(6) Bring forward any planned sales of capital assets.

(7) Increase sales volume or price.

The efficient and effective organic farmers who were interviewed in connection with this book all had a very clear idea of the monthly cash flow. Methods of improving the availability of cash are listed below.

When money is coming in too slowly
If this happens:

(1) Invoice more promptly.

(2) Invoice at the end of every month. If you fail to do this, the invoice will be treated as next month's invoice, and payment will probably be delayed by another 30 days.

(3) Put shorter terms of payment on the invoice. The debtor may choose to ignore this, but at least he may then be contacted earlier. This creates pressure, which should on the whole result in earlier payment.

(4) Chase bills, either in person, by telephone, or by reminder letters

and statements. (It is advisable to try to obtain credit references on your customers before supplying large orders.)

(5) Persuade the buyer to pay by standing order if he is a regular customer.
(6) Persuade the buyer to pay immediately on delivery.
(7) Persuade the buyer to pay in instalments.
(8) Put a percentage on the invoice if not settled by a certain date, or offer incentives for prompt payment.

When sales are down
When this happens, try to find out why. Is it due to poor sales efforts, a drop in quality of the product, pricing, increased competition, production problems or a combination of the above? Or is it that the volume is not high enough? Readjust your cash flow forecast and decide what to do to improve sales. Although it is rarely possible to vary output at short notice in many farming systems, it can be achieved in beef production where animals can be sold almost at any stage of their life. The farmer must establish whether revenue can be increased more easily by raising price or by trying to sell more goods at a lower price level. The relationship between price, demand and revenue will be different for different products.

When the farmer is paying out too fast
It may be possible to obtain more favourable terms from creditors. This should be done with caution, and it is sensible to pay the creditor something each month. If this is impossible, then it makes sense to contact the creditors and agree a plan for future payment.

If at some stage the cash flow is better than anticipated it is sound policy to keep a surplus against possible future negative balances. Alternatively, it is possible to increase stock numbers on the enterprise that the costings exercise has shown to be most profitable. In farming, this can be achieved only if the season is right. For example, wheat acreage could be increased only at the time of ploughing and if it fitted in with the planned rotation, whereas extra sows could be brought in, almost at any time, if the pig enterprise was flourishing.

Cash flow during conversion process
The monitoring of cash flow is particularly important during conversion from conventional to organic farming. Certain systems may produce no returns for a year or more, suckler beef, for example, and under these circumstances it may be necessary to explore other

methods of income, such as bed and breakfast, camping, horse riding or motocross. An alternative is to reduce expenditure by not converting the whole farm immediately, especially on the larger farms.

On some farms which have a considerable proportion of conventional grassland as temporary leys in an arable rotation, the decision to delay converting to organic cereals before converting to organic beef and sheep may leave the livestock short of grazing. In this instance the arable and the temporary grass should be converted before the beef and sheep. However, care should be taken with the piecemeal approach to conversion. For instance, having an organic flock of sheep and a conventional one may make such exercises as dipping and feeding during housing very complicated.

Costings

Carrying out a costings exercise over a fixed period provides information needed by the farmer on recent production costs; this can be used in planning and decision-making on issues such as:

- Pricing and output, and calculating the break-even point.
- Financial performance of the various farm enterprises; the method is to allocate costs to each separate enterprise, so that these can be compared with sales figures and also show how well each enterprise has performed. Are the pigs doing better than the sheep, and what about the hens?
- Whether to pursue a new activity, for example, setting up a B&B, or adding a children's play area for visitors.

The exercises described below will help the organic farmer to decide whether an enterprise should be started, expanded, modified, or even discontinued. There are different methods of costing: absorption or full costing, and marginal or variable costing. These have different uses which are discussed below.

Absorption costing
In this exercise everything is costed, including variable (or direct) costs such as labour, animal feed and seeds, as well as fixed costs (or overheads) such as rent, buildings, machinery, depreciation and even factors such as the time spent trying to market produce. Fixed costs,

unlike variable costs, are those whose levels do not alter when production levels change.

Figure 8.2 illustrates the format of a typical absorption costing exercise. This exercise can easily be used to calculate the overall profit or loss of each enterprise and its profit margin. See for example Fig. 8.3. One of the difficulties of this exercise is allocating the fixed costs between the various enterprises. Take housing for example: if the shed is shared between beef cattle and sheep, then the allocation of the cost could be made on a space and time basis. On a large-scale farm with several workers there will probably be a shepherd, a stockman, and a tractor driver for the arable crops. However, on a smaller farm where the farmer and his family do all the work then division of time between

		Enterprises		
	Wheat	Sheep	Hens	Total
1 Direct costs				
Contracted labour				
Feedstuffs				
Seed				
Vet and med				
2 Overheads				
Depreciation				
Machinery				
Buildings				
Marketing costs				
Farmer's time				
3 Total cost (1+2)				

Fig 8.2 A typical absorption costing exercise.

		Enterprises		
	Wheat	Sheep	Hens	Total
1 Sales				
2 Total cost				
3 Profit/(loss) (1–2)				
4 Profit margin (3/1)				

Fig. 8.3 Calculating profit/loss and profit margin.

enterprises is harder, and the farmer and his family must be paid a notional wage so that their time input can be costed. If this notional wage is fixed at a low value, then the farm may appear to be making more money, but at the expense of the family's standard of living.

Marginal costing

This method of costing differs from absorption costing as it does not include fixed costs. This means that it is relatively easy to calculate on a product-by-product basis since there is no need to make arbitrary judgements on the allocation of overheads. Its value is that it can show the minimum price that can be charged for each product, which will help with selling. Marginal cost for an enterprise includes all the direct expenses, such as feed costs, that vary when the level of production changes. It is defined as the change in these direct costs as production increases by one unit. That is, how much costs would rise if the farmer bought one more cow or sheep or produced one more tonne of wheat, for example. It can show the effect of altering factors such as the number of cows, the level of concentrates fed and the stocking rate. Figure 8.4 shows the format of a typical marginal costing exercise, but again this is not intended to be an exhaustive list of cost items. As before, this information can easily be combined with sales revenues to show the contribution (or gross margin) made by each enterprise towards fixed costs (Fig. 8.5).

Marginal costings deliberately exclude fixed costs, but it must always be remembered that these still need to be paid. It is important to think about the difference between marginal cost and marginal

		Enterprises		
	Wheat	Sheep	Hens	Total
1 Contracted labour				
2 Feedstuffs				
3 Seed				
4 Vet and med				
5 Total direct cost (1+2+3+4)				
6 No. of tonnes, animals, acres, etc				
7 Marginal cost per unit (5/6)				

Fig. 8.4 A typical marginal costing exercise.

	Enterprises			
	Wheat	Sheep	Hens	Total
1 Total sales				
2 No. of units sold				
3 Revenue per unit (1/2)				
4 Marginal cost per unit				
5 Contribution per unit (3–4)				
6 Total contribution (2 × 5)				

Fig. 8.5 Calculating gross margin.

revenue (revenue per unit) as a contribution to fixed cost. In dairy farming it is common to express the contribution as gross margin over feed costs, leaving out other variable inputs, as the feed costs represent the greatest proportion.

In summary, there are three possible results from this exercise:

Negative contribution per unit
In this case, the marginal costs per unit are greater than marginal revenues (i.e. selling prices), so the farmer will be in trouble. He or she would effectively be paying customers to take away the produce, and making a loss on every sale. This should not be sustained.

Positive but insufficient contribution
Here, marginal costs are lower than marginal revenues, so a profit is being made, but the level of contribution is not enough to maintain fixed assets and to leave sufficient profit to make the business worthwhile. This can be sustained for the short-term only, unless the enterprise is in some way contributing to the success of a more profitable product.

Positive and sufficient contribution
Marginal revenue exceeds marginal cost, and the total contribution is enough to cover fixed costs. This can be sustained.

The distinct advantage of marginal costing is that it helps the farmer to realise the absolute minimum he has to earn on his products. This is of special value when the farmer is selling directly at the farm gate.

Using marginal costings, he can work out the cost of producing a dozen eggs and then add a 'reasonable' profit margin. 'Reasonable' is best defined as a price that the market can bear and yet one that helps the farmer recoup costs.

Recently, organic farms have been encouraged to make added value products such as cheese and yoghurt. The farmer processes the milk and can fix a price for the farmhouse-produced cheese or yoghurt. It needs to be a unique quality product, or have a superior image, if it is going to sell well at a higher price than a similar product in the local supermarket.

Costings and decision-making

Keeping good records for each product produced allows the farmer to assess the profitability of each enterprise. If an enterprise has been shown from the costing exercises to be consistently losing money, then it may make sense to scrap it. However, before doing this various other factors should be considered:

- Whether the husbandry of the production system is being efficiently carried out, and how much scope there is to increase reproductive efficiency in the sheep flock, for instance, by reducing the number of barren ewes and by reducing the number of lambs lost before sale.
- Whether this enterprise contributes to other enterprises: for instance, farmyard manure from the housed beef animals in winter may be essential for the continuing fertility of the silage fields, and straw from the wheat may make a useful contribution to the cost of bedding for the cattle; alternating beef and sheep grazing will help to reduce parasite infection for both species (Newton, 1993).
- An attempt should be made to decide whether the price structure of the enterprise is likely to improve in the future, by consulting reliable advisors. The profitability of pigs and turkeys is well known to be cyclical; a bad year is therefore likely to be followed by a good year.
- Before scrapping an unprofitable enterprise it is also sensible to try to determine whether financial help from the government is likely to be given in the near future.

The costings exercises can also help with decisions on pricing (see Chapter 9, Marketing Organic Produce). The absorption costing exercise shows the minimum level at which the products must be priced in order to cover all costs and give a reasonable profit margin. The marginal cost exercise shows the absolute minimum price that the farmer can afford to sell at in the short-term. It is important to note that pricing at marginal cost is only a short-term tactic, for an emergency. Continuing it long-term will be akin to bleeding the business to death slowly. You will have no money for repairs or replacements or for your own retirement.

The costings exercise also shows the output that is required to reach break-even point. The break-even point refers to the price and quantity sold that will just cover all costs. At this price and sales level, no profit is made, nor will you be losing money. Break-even is when:

$$\text{Selling price} \times \text{Quantity sold} = \text{Fixed costs} + (\text{Variable costs} \times \text{Quantity sold})$$

The break-even point is calculated as follows:

$$
\begin{aligned}
\text{Sales price} \times \text{Quantity sold} &= \text{Revenues} - \text{Variable costs per unit} \\
&\quad \times \text{Quantity sold} \\
&= \text{Contribution margin} - \text{Fixed costs} \\
&= \text{Zero (or substitute a desired profit margin)}
\end{aligned}
$$

By varying the prices, quantities and costs, the farmer can assess the potential profitability of the different products or enterprises. One approach is to estimate how much has to be sold in order to break even at a particular price. Alternatively, the farmer can estimate how much is available for sale and, on that basis, work out what price needs to be charged in order to break even. If the price appears to be too high, it is for the farmer to decide between increasing sales or reducing costs.

When it comes to the purchase of expensive indivisible items such as the machinery for making silage, then, for the smaller farm, consideration should be given to sharing machinery with neighbouring farmers, or subcontracting to specialists. Machinery is expensive to purchase. It can take a long time to recoup £50 000 for a machine that is used for only a week each year. The cost of hiring a contractor to combine and harvest cereals was calculated to be £96 per ha (Lampkin & Measures, 1994). With 100 ha of cereals to harvest, this is less than

£10 000 per annum. Sharing machinery with neighbours can sometimes be exasperating when it is your turn to be last, especially when the quality of your hay is deteriorating, but it may be the only solution you can afford.

References

Lampkin, N. & Measures, M. (1994) *Organic Farm Management Handbook*. University of Wales, Aberystwyth and Elm Farm Research Centre, Newbury.
Newton, J.E. (1993) *Organic Grassland*. Chalcombe Publications, Maidenhead.

Further reading

Hingston, P. (1988) *The Greatest Little Business Book*. Hingston Associates, Perthshire, Scotland.
Rosthorn, J., Haldane, A., Blackwell, E. & Wholey, J. (1986) *Small Business Action Kit*. Kogan Page, London.

9   Marketing Organic Produce

The organic farmer has to do more than merely farm, following organic principles, in order to be successful financially. Produce has to be sold advantageously not just once, but on a regular basis, every month, every year. Where the farm is one of the first in a particular region, it may well be the farmer who has to create the distributing, marketing and processing infrastructure. Until a smooth mechanism for selling organic food develops in that particular region/country, the organic farmer will face relatively high costs for segregating organic and conventional products in storage, handling, packaging and processing. 'Until a sales infrastructure is developed, so that the organic produce can move from the farmer to the retailer in a smooth and flexible way, it will be expensive and difficult' (Wier & Calverley, 2002). Some farmers may sell directly to the consumer, some may use intermediate channels and others may use a combination of methods. Some farmers may decide to sell their raw produce whilst others may choose to process the product themselves.

These individual organic farms have much in common with small and growing businesses in other industries. The farmer as owner/ manager will have to juggle a variety of roles. These will include organic production, market research, targeting markets, advertising, going out and selling, managing finances, and dealing with staff, customers and day-to-day operations. The size of the potential market will also have to be assessed; and considerable time, effort and money will have to be spent in order to establish organic products in the market. The farmer will have little clout when negotiating with powerful buyers, suppliers or traders, and will have to compete with other players already in the established market. It is important to bear in mind throughout that success needs to be as fully analysed as failure, so that it may be replicated in a variety of circumstances.

Challenges such as these, in the developing market, are discussed in this chapter, and possible solutions put forward. It is important to note that the rate of development of the organic sector differs from country to country, that some inspection bodies are more trusted than others, and that each country or region will have its own peculiarities.

Challenges and opportunities

Big corporations

Large supermarket chains, food processors and manufacturers have become involved in organics. These outlets provide good publicity, as well as new outlets and distribution networks, but they also present new competition, by producing and marketing their own organic range. Organic supermarkets have been set up in the United States, stocking not only organic products but also other natural health and special dietary products. Distributors and wholesalers are following suit, in Britain as elsewhere: organic eggs are now loaded onto the same truck as free-range eggs, Freedom Foods eggs and eggs from rare breeds. As with the UK, it is significant that many US manufacturers and distributors, including multinational corporations, are specialising in processing and marketing organic products. Long-time manufacturers of organic foods have also introduced new organic lines to their product ranges.

There is pressure from some produceers and consumers to increase the level of processing. The involvement of large supermarket chains has increased the scope for processed products. These organisations are likely to source from overseas markets if it is advantageous. 'This trend towards industrialization of organic agriculture is of course resisted by many in the organic industry who see it as a move away from traditional values. The size and nature of the certified organic market in the future depends somewhat on how this issue is resolved' (FAO, 2002).

Direct markets

Farmers' markets have enjoyed a small revival, especially in the USA where these activities are fostered by state and local municipalities. Some states have even organised organic-only farmers' markets and developed logos to promote locally grown food (Dimitri & Greene, 2002). This resurgence of farmers' markets may also be due to consumers enjoying the feeling of obtaining their food directly from producers.

Changes in organic farm size

Traditionally, most organic farms have been family farms. They were usually small, low-tech, eco-friendly, selling locally. However, some very large-scale farms in the UK, USA and Australia are under conversion. For example, Pavich Family Farms in California has over 1600 ha of organic land and 200 ha in conversion. It supplies the market with table grapes, and other fruit and vegetables (FAO, 2002). In Britain, Red Kite Farms is currently one of the biggest organic dairy businesses. It is claimed that it has already taken over a contract to supply a large supermarket chain, a contract which was previously held by a collective of organic dairy farmers. Australia has about 7.5 million hectares of land devoted to organic farming, most of which is dedicated to organic beef and lamb enterprises (FAO, 2002). Growth in the Australian industry has been stimulated by strong overseas demand for organic produce.

Current market channels and products

Currently, 74% of all organic food in the UK is sold via the large supermarket chains, the rest being equally divided between direct marketing methods and independent stores (Soil Association, 2003). In the USA, organic foods are currently sold through a wide variety of outlets: until 2000, most retail sales were through health and natural food stores (68%), followed by direct markets, and only 7% through conventional supermarkets; by 2000, 49% of all products were sold in conventional supermarkets, 48% in health and natural food stores, and 3% by direct methods (Dimitri & Greene, 2002). In 2000, organic produce was available in 73% of all conventional American grocery stores. In Sweden and Denmark, 85% of organic food is sold through conventional retail stores, the rest being sold by direct methods.

In the USA, desserts made up the majority of new organic products in 2000, while the highest for 1999 were beverages. Fresh produce is the top-selling category, followed by non-dairy beverages, breads and grains, packaged foods (frozen and dried prepared foods, baby food, soups and desserts) and dairy products. Thompson (1998) stated that the market share of organic baby food is larger than the share of all other organic food in the USA. Organic baby food is also in demand in the UK market, and Baby Organix, a UK brand, has even brought out organic prepared meals and snacks for toddlers. However, the purchase of organic produce by these consumers seems to tail off after this. It would be interesting to follow the buying behaviour of people who bought organic baby food and to find out how long this buying behaviour lasts. Most of the organic foods in UK are bought by households with no children (Cooke, 2002).

Organic products in Germany, France and the Netherlands are sold mostly through health-food stores. This is not surprising, as shoppers in these countries tend to prefer buying from independent butchers, grocers and fishmongers, to supplement discount grocery purchases (Brassington & Pettit, 2003). Their supermarkets have a strong discount retail culture. On the other hand, consumers in the UK and USA prefer a one-stop-shop with everything under one roof. The dominant position of the supermarkets means that their labour costs can be 10% to 20% lower than the independent grocers, and their buying advantage 15% better. In the UK, large supermarket chains are dominant, and this is where most consumers buy their organic food.

Austria is the largest organic producer in the EU, most farmers being feed producers. It exports a third of its total organic output. In the early days, most of the produce was sold through farms and farmers' markets. In 1999, the two largest supermarket chains sold 90% of the organic produce under their own brand names. The rest was sold directly by the producers and in special health food stores. The first organic supermarket, 'Bio market', was opened in Vienna in 1999 (Krucsay, 1999).

At the moment, the UK imports about 75% of its organic food. Germany imports 50% to 60% of all its organic products (Wier & Calverley, 2002). Market share of the organic food for Germany, Sweden, Austria and Italy has reached between 1.5% and 2.5% of the total food consumption. Denmark currently has the highest per capita consumption of organic food in the world, with organic food representing about 3% of all food consumption. The organic dairy sector is said to have reached maturity: up to 80% of consumers buy organic dairy products, and a third of all milk sold is organic. There was an oversupply in 2001, with half of all organic milk being sold as conventional milk. The export market seems lucrative, with the USA being the latest importer of Danish organic milk.

In Denmark, the organic food market was initiated by government subsidies and advisory services to organic farmers. Until 1993, the consumption of organic food was at a steady 1% to 2% of the total food market share (Wier & Calverley, 2002). However, consumption escalated when supermarkets lowered the prices by 15% to 20%, increased supply, and initiated intensive marketing of the products. With support from the Danish authorities as well as the close co-operation between farmers, producers, manufacturers and retail trade, the following years' annual growth rates were about 50% to 100% for most products (Organic Denmark, 2003).

New figures from GfK, a Danish market research agency, show that total market share for organic meat, fruit and vegetables, in the first 6 months of 2003 was 5.6%. Doorstep schemes are attracting interest. Some examples of the activities organised:

- A retail chain had promotions throughout spring and had new displays, posters and display boxes. They also launched a new product, the extra low-fat Jersey milk, stimulating further interest. Organic Denmark (2003) ran campaigns for dairy products, meat, fruit and vegetables. They assisted with product development, marketing and in-store exposure.
- Eight supermarket chains, together with their meat processor and marketing companies, each held week-long campaigns, with special offers for organic meat. These were advertised in their circulars. This resulted in a 23% higher turnover than in the same week of the previous year.
- A Danish co-op, in collaboration with the Danish Organic Service Centre, organised two organic 'caravans' to visit 70 of its stores for a week at a time. The aim was to provide information about organics, sample organic food and to meet an organic farmer. The Organic Service Centre arranged for a local organic farmer to join the caravan at each store. They also manned the caravan with their own trained employee, to provide information on organic produce. Samples and leaflets were given out. The co-ops provided a huge toy cow which was manoeuvred by a young 'demonstrator' who handed out cookies. The campaign was reported to have boosted sales by an average of 50% compared to the same week in the previous year. The Danish government covered 50% of the costs of the promotions, with the retailer and organic producers paying the other half.

Organic symbol

Consumer studies suggest that trustworthy labels guaranteeing organic production are very important for consumers (Wier &

Calverley, 2002). The certification bodies should do much to improve their recognition by the public if they are to help farmers sell their produce. The organic symbol has itself to be a strong market icon. If not, consumers will find it hard to separate pseudo-organic and 'environmentally friendly' products from authentic organic products. There is only one accrediting label for organic production in Denmark. It is controlled by the state, is well known by 50% to 75% of all consumers and is apparently greatly respected.

Yet it is also important to realise that simply having the organic logo on the product may not be enough to sell it: production and processing features simply may not be enough reason for the consumer to buy. For example, two major trends in the industrialised world are the demand for convenience food, and health consciousness for healthy eating patterns (Cooke, 2002; Wier & Calverley, 2002). Taking this into account, the farmer might consider packaging organic poultry with a packet of herbs/spices, thereby conveniently providing everything for cooking. Alternatively, the meat could be cut up into easy-cook pieces, each with its own set of prepared ingredients.

Price

In working out a reasonable price for produce (see Chapter 8), the farmer should be aware of current and historical prices, and also of predictions. By subtracting a percentage margin from the going retail price, it is possible to estimate the wholesale price.

The organic farmer also has to take into account the charges made by the organic licensing bodies. In the UK, charges differ for the various licensing bodies and in most cases are based on selling price or acreage, with no specific maximum. Furthermore, a levy has to be paid at every stage. For example, at present, levies must be paid on beef at five stages: the farm, the abattoir, the processing or packaging plant, the wholesaler and the retailer.

Consumers seem to be more influenced by absolute price than by the price differential (Jones, 1989). The view is that conventional fillet steak is already expensive and anything added because it is organic makes the price prohibitive. However, a good premium is available on mince, which is a cheap product. It may also explain the popularity of repackaging organic eggs into a package of only 4 eggs, instead of the usual 6 or 12.

In the USA, frozen organic vegetables, organic milk and organic baby food all exhibited high price elasticity of demand: that is, quantity purchased responds greatly to price changes (Dimitri & Greene, 2002). It appears that a lower price premium would induce a considerable proportion of consumers to buy organic products. The consumer considers price to be one of the most important factors when buying organic food (Cooke, 2002; Wier & Calverley, 2002). This is further borne out by the experience in Denmark, where the consumers show a very low concern for health risks in their food consumption and yet are currently the highest consumers of organic products. Consumption escalated when Danish retailers lowered the prices by 15% to 20%, increased supply, and initiated intensive marketing of the products.

Market research and assessing your market

Before beginning to farm organically, it is important to do as much as possible to ascertain the best market for your likely organic products and to marry this with the type of farming that you want to do and the production systems which suit the farm you want to either buy or convert. The 'best' market can be defined as the one that will provide the highest and most regular income.

Desk research
Desk research consists mainly of reading as much as possible about other people's experiences and recommendations. Read industry publications to find out what is currently in demand and what looks promising for the future, as well as what price ranges are available for various products. If there are one or two people who have published material of particular relevance, then if possible go and speak to them in person. They may say more in person than they will write, or their work may prove to have been derivative, in which case find the original source. Market research reports and trade publications can be obtained from most local libraries. Most broadsheet national and local newspapers report occasionally on the organic food industry, and sometimes it is possible to identify a gap in the market.

Talking to potential buyers
Potential buyers may be intermediaries such as distributors, whole-salers, processors (for example, millers, meat-packers and cheese

makers), local shops or supermarkets, depending on the proposed product. They will advise on the particular lines they require and will discuss quality, quantity, regularity of supply and pricing structures that are available. Buyers are one of the best sources of information for the producer, as they have regular contact with the final consumers and they can also provide valuable knowledge concerning other farmers and trends in the market. Attending relevant trade shows is often a way of making contacts in the industry.

Another means of obtaining information about the market would be, for example, to speak to hotels, health clubs, sports and recreational clubs, corporate offices, restaurants, pubs, airlines, schools, hospitals and other institutional buyers, to find out whether they would consider buying a regular supply of organic produce, such as eggs, meat, and meat products. A few nurseries/child care centres in the UK use organic produce when preparing meals for the children in their care. They use this as a selling point in their promotional material. Another example comes from Austria, where 28 Viennese hospitals carried out a pilot test of organic foods (Krucsay, 1999). Private caterers, chefs, food critics, nutritionists, and journalists from food publications are both a source of information and a means of encouraging sales. Introductions by a mutual business associate are particularly useful in this context.

Speaking directly to the consumer

A number of academic and industry research papers into consumer behaviour and motivation for buying organic food are now available. The results so far have been inconsistent and, at times, contradictory. However, this may be due to the different methodologies used.

The principles of marketing organic food are the same as with conventional foods. There may be different sales and distribution channels but the factors influencing consumers' buying behaviour will be similar. Therefore, the farmer thinking of selling direct to the consumer will have to sell the benefits of a particular product, and these can include health, nutrition, taste, image, lifestyle, environmental concerns, and convenience. The farmer has to find out about the preferences of the target group, in terms of packaging, display, pricing and service, and must also be aware of what the consumer buys and eats, where and when they shop, age, ethnicity, religion and attitude to life.

Consumers who participated in a survey for the Leopold Center for Sustainable Agriculture in the USA (The New Farm, 2003) preferred locally grown food over organic, provided that price and appearance were the same. Over 75% of consumers chose produce 'grown locally by family farmers' as their first choice. They were most responsive to labels that connected product freshness with the time (in days) that it took the produce to travel from farm to store. However, in the larger supermarket environment, consumers do not seem too concerned about the origin of the products (Roos, 2002).

Final decision

If the above exercises have resulted in one or two promising lines, the farmer must then examine his own ability and stamina realistically, bearing in mind that good communication skills are likely to be a prerequisite for success. If the farmer has reservations regarding motivation and the effort involved, further research is advised.

If the research as to likely organic products that will sell comes up with a series of negatives, then the farmer has to fall back on his own creativity. As Finke (1989) pointed out, he could not persuade any bakery in the nearby towns in Ireland to bake organic bread from his flour; they said there was no market. However, he finally persuaded one bakery to bake it, on contract, provided he did the selling. They agreed, he got the packaging printed and after five months they had sold 10 000 loaves to various shops in the town. After the consumer preference had been demonstrated, there was no longer a problem with the bakers.

Finance

It is important to remember that new customers are expensive – they need a great deal of convincing – whereas existing customers may simply need continuing confirmation of brand quality.

It is essential to collect sales figures and the associated costs by the day/week/month, and to separate the figures for the different products. It is also essential to distinguish between fixed and variable costs. As we have seen (Chapter 8), the latter include costs such as field preparation, seed, harvesting and packaging. Fixed, or overhead, costs include loan repayments, property taxes, insurance, depreciation and maintenance on buildings and equipment. It is important to include the farmer's salary as a fixed cost, as well as marketing costs, deliveries, fuel, and vehicle upkeep.

Outlets

It is possible to sell directly to the consumer (via farm shops, farmers' markets, mail order and internet), or to restaurants and institutional buyers, or to brokers, distributors, wholesalers, processors, food manufacturers, millers, abattoirs, butchers, farm shops, co-operatives, local retail stores, speciality stores and large supermarket chains. If one outlet does not work, it is possible to switch to another. Some farmers use multiple outlets, selling directly to consumers as well as commercially to middlemen. In this way, the farmer is not reliant on just one outlet. Some farmers also take the decision to have their own processing facility, such as milk processing or cheese-making.

Farmers' markets

A farmers' market is where a fee is paid for a market table, on a particular date. Here, the farmer displays organic produce and keeps all the proceeds. Some farmers have made contact in this way with fastidious restaurant chefs in search of premium ingredients. An ability to answer relevant questions helps in this context, as do tastings and, if possible, simple demonstrations. A quick walk round a farmers' market is also a simple way to compare products and prices.

There has been a resurgence in farmers' markets in the USA. There are several advantages in this method of selling: for example, more customers are attracted due to the variety of products being sold at one place; and some customers are able to source special items in large quantities for home or their organisations.

However, the disadvantage is that price comparison is easy and buyers in farmers' markets are more bargain conscious than customers who visit individual farm locations. Also the farmer will have to take time away from production operations on the farm.

The farmer should price his product to reflect the quality, cost of production, competition, service provided, convenience and types of buyer targeted. 'If all of it sells, the price is too low. At least 10% of people should walk by shaking their heads.' The farmer has to consider the transport needed for his product to the market, the distance involved and whether a suitable vehicle is available and can be fully loaded. Sales must be assessed after every event. 'Don't drive 200 miles to sell 25 chickens' (Fanatico & Redhage, 2002).

Farm shop

The farmer would need to enjoy meeting people, as well as dealing with the other retailing skills of controlling quality in the shop, arranging displays, adjusting prices, packaging and merchandising.

The most important points to consider regarding the farm shop are the following.

Location

To get the necessary volume of produce sold it helps to be in a busy area. This means either being near a large town or on a busy road, but not so busy that people in cars are unable to stop. Advertisements and an occasional press release are useful tools in reminding the public of the shop's location and its range of products. In the UK, 30 miles is generally regarded as the maximum people will travel in order to purchase specific goods. Parking facilities and liability insurance must be in place. The shop may be combined with holiday accommodation, gift shop, a restaurant or children's play area.

Quality

The products should always be of high quality. This is even more important with organic than with conventional food; the organic apple has been derided as the one with the caterpillar for too long. If the produce is found to be substandard by the consumer, he or she may come back for a replacement once, but not twice.

Variety

It is important to be able to offer customers a wide variety of quality organic products. As it is unlikely that these can be produced on only one farm, contact should be made with producers of other organic products. Thus a range of organic products should be offered, such as meats (lamb, beef, burgers, sausages, pies, ham, bacon), eggs, dairy products (milk, yoghurt, cheese), bread, biscuits, fruit, vegetables, wine and so on. Identify the most frequently bought items, and display them in such a way as to encourage the customer to visit all parts of the store. Place some of these items close to the store entrance.

Packaging

The presented produce ought to look clean and wholesome and consistent with the organic message. For example, the farmer may be creative with the packaging, using recyclable brown paper bags to achieve a rustic look, or a more sophisticated design to create a more upmarket image. It is important to use a trademark with known qualities, for example, the Soil Association symbol.

Staffing

The bigger the farm shop, the more staff are needed to deal with the customers, and the more stock control, insurance and theft protection become essential. It may become a big enterprise which would then require sophisticated organisation. Is this what you want, and is this what you are good at? If not, be careful.

Cash flow

A daily cash float is required in the shop. It is essential that takings are totalled up daily and accounting records are kept accurately.

Environment

The Food Safety Act 1990 applies to anyone owning or managing a business producing food for human consumption. The food hygiene standards that are required are described in the publication *Success with a Small Food Business* (1994). Key areas are the cleanliness of the food being sold, the use of easily-cleaned, well-ventilated and properly lit premises and a high standard of personal hygiene. Fish and fish products, meat and meat products, milk and milk products and egg products all require particular care, and many products require temperature control at all stages of production, distribution and display.

The regulations as to what may be labelled 'organic' are stringent, particularly where the product contains multi-ingredients with less than 100% organic content. In 1995, at least 50% of ingredients of agricultural origin were required to have been derived from products produced to organic standards by an approved producer holding a valid certificate of registration.

Attracting people to the shop

It is worth considering signposts on the road, and leaflets and posters at local shops, garages, tourist information centres and post offices. The signposts should be well-maintained and give enough warning to drivers, so that they can pull up in time. Advertise locally, perhaps using radio stations or newspapers. Information may also be conveyed via the newsletters produced by local clubs or associations. Cultivate goodwill with the customers, attract people back and try to build up a regular clientele.

Educating the consumer

The farm shop proprietor may need to explain to customers why organic products are different or superior. For example, many consumers will not be aware that organic meat is likely to be slightly darker and more dense than conventional meat. The fact that it has no water or colouring added will affect its appearance. Beef and lamb raised entirely on grass are not as heavily marbled as grain-fed beef and lamb. Seasonal variations in the availability of fresh products should be presented as a virtue and proof of freshness. Similarly, variations in size and shape may be seen in a positive light.

Even the provenance of products such as potato crisps, or the effect of recent weather conditions, may be of interest to some consumers. For information regarding nutrients, retailers may wish to refer consumers to the relevant literature.

Mail order

This comes under various names such as box schemes, home deliveries or doorstep deliveries. This sales method involves the potential customer purchasing the product from an advertisement, a catalogue or a web site or even from the retail store itself. Several butchers have run successful schemes via their retail stores.

Administration

It is necessary to be strong administratively. The customer is being dealt with by telephone and by letter, and the person or persons responsible must be available full time and be courteous, efficient and well-organised.

Promotion

A mail order business needs to be promoted. This can be achieved by attending shows, using flyers and by advertising. People soon forget advertising, unless they happen to be looking for the product that is being advertised at that precise moment, so just one advert may not be enough. The rule of thumb is 4 to 5 hits. Alternatively, it is possible to send out a press release to relevant publications. This is usually a newsworthy item which the reporter may wish to include in an interesting article. Reporters will usually read the article over to the author before publication, but the editor has the final word. It is essential in mail order selling to have a catalogue of products with prices, and a form on which customers may fill in their address and details of how they will pay.

Delivery

In dealing with orders, it is necessary to decide on the method of delivery, whether by post or by your own transport. If you do not impose geographical restrictions, then the cost of extra transport must be included in the price. The item requested must arrive promptly and in excellent condition.

Payment

Check whether customers placing large orders are creditworthy, get payment first, and decide in advance whether to accept credit cards and/or cheques.

As with all successful selling, the interchange with the customer and the mutual trust that can be built up are vital.

Many of the successful schemes have involved fruit and vegetable growers. For the meat producers, it is still risky. Consumers sometimes expect to pay less since they are buying direct and in bulk. A large number of sales means more time is required for order processing and delivery, to generate the same amount of sales.

It is now realistic for even the smallest company to develop and manage some kind of database relatively cheaply. It is possible to hold a considerable amount of information on each individual customer, and it is relatively quick and easy to update the data and use the information automatically (for example, on invoices or address labels).

The challenge is to communicate actively with the customer and continue to make sales after the first order. It is important that the farmer keeps in regular contact with the customers, and develops other incentives to ensure they keep ordering.

Farmers can develop their mailing lists from contacts made at farm shops and farmers' markets. They can also buy mailing lists or obtain lists from buyers' clubs and from exhibitions and fairs. They could invite consumer groups to visit their farms, making sure the visitors leave a contact address for future promotions.

Internet

The internet may be used as a form of mail order or purely for advertising. Farms may either maintain individual websites or participate in a directory listing. Costs and providers need to be researched, and it is possible to barter for a website design. Customers like a website that is easy to use, quick to download and updated frequently. Avoid graphics that take a long time to come up on screen.

It is important to check whether existing customers are on the internet, or have e-mail, and to be aware of the barriers to internet buying: pricing (transportation costs, including items returned), credit card concerns, privacy issues, navigating the site. Customers may be reluctant to make payment over the net, and should be encouraged to purchase over the telephone after browsing through the website. It is important to provide details of *all* the ways through which the vendor may be contacted.

Restaurants, kindergartens and other catering outlets
This includes schools, hospitals, health clubs and airlines. The farmers may either approach the buyer within the organisation or the caterer tendering for the contract. Building a strong relationship with the buyer or chef is vital in this market. Keep the buyer informed of any prediction of a shortfall or oversupply, or variation in quality. This allows them to take other action accordingly.

The farmer needs to consider the cost of delivery to each eating establishment. By selling directly to restaurants, the farmer may achieve a higher wholesale price but there will also be a need for high-level service to each customer; and there will be stiff competition from wholesalers who have a year-round product line and sales staff.

The farmer and restaurant may work together to promote the produce used. For example, the restaurant could provide leaflets containing information on the produce, farm source and animal welfare.

Meat producers often experience difficulty co-ordinating the complex management of production, processing, delivery and sales system required to target the restaurant market. Compared with wholesalers, individual restaurants do not use large quantities of meat. It would be advantageous for the farmer to be near a large metropolitan area with numerous restaurants. Access to a variety of restaurants will allow the farmer to use up more of his animal. The norm is for the farmer to establish a route and deliver once or twice a week. The farmer could also sell meat bones to chefs who appreciate the quality for soup stock.

Some restaurants and most institutions (for example, schools) have long-term contacts with suppliers, with the institutions having more layers of bureaucracy. They also usually ask for volume discounts, so the farmer will have to adjust the initial pricing accordingly. Some institutions contract out their food to caterers and this may or may not be an advantage to the farmer. If a buyer is interested in organics, this may help sell the idea to the institution. In some cases, buyers are not interested in fresh, local meat, and prefer frozen, pre-cut and even pre-cooked.

Success in this sector greatly depends on the personalities and relationship between the parties, as well as being able to offer consistent supplies of quality produce. When approaching restaurants and institutions, it is important to provide them with promotional material and even samples. The farmer may need to work with the chef to discuss whether different cooking methods may be required. For example the low-fat content in organic meat means that product yield is higher (Born, 2000). The Organic Valley dairy co-op based in Wisconsin, has managed to get schools to offer organic milk, via vending machines. They are also currently developing 'kid-friendly' cheeses and strawberry and vanilla flavours of single portion milk (The New Farm, 2003).

Finally it is important to use a receipt book with duplicate copies and to get someone to sign it when a delivery is made.

Distributors, brokers, wholesalers

Distributors and wholesalers may act as middlemen, moving raw products from the farmer to the processor or manufacturer, or moving processed food to retailers. They will have information on how much is needed by the manufacturer and what quality. Brokers may also do the marketing and check that stores properly present your product.

The farmer should keep the distributor/wholesaler informed of production plans, and must give the buyer as much notice as possible of any change in plans. This will provide an opportunity to line up another supplier (Roos, 2002). Some wholesalers offer farmers long-term contracts which guarantee price premium for organic products over several years.

Food processors and food producers, including millers and butchers

The farmer should find out as much as possible about the suitability and experience of the processor. It is important to be aware of the level of quality that is achievable, products produced for other clients, financial status, paying behaviour and integrity.

The farmer can sell to the processor, who then sells it on. The farmer could also have a partnership agreement with the processor to sell to a certain market. There should be a decision as to whose name/brand it will be marketed under, the producer's name or the processor's name.

Alternatively, the farmer can also contract the processor for his processing service, leaving the farmer to market the product himself.

Food products may include frozen produce, canned produce, pasta, sauces in jars, prepared meals or pizzas. It is important to work closely with the manufacturer, in working out the best types of ingredients to use, depending on the manufacturing facilities available. For example, frozen fruit and vegetable processors require a specific size and quality, pasta and bread manufacturers will have their specific requirement of the variety and grade of wheat. They require consistent quality and volume so as to be able to produce a consistent end product. As with restaurants, it is important to build a good relationship with your processor and to keep the manufacturer informed of any potential problems. This allows them to take other action to ensure their production does not suffer.

Two major trends in the industrialised world are the demand for convenience food and awareness of the need for healthy eating patterns. The trend is away from generic foods towards processed products, for example, as we have seen, there is a growing demand for organic frozen desserts and items such as frozen pizzas. Many supermarkets see a potential for surplus products in freezing and canning surplus produce (FAO, 2002). The popularity of functional or healthy foods is also a potentially lucrative area for organic produce, especially dairy products.

Millers
Marriage and Starling (1989) both advocate the signing of forward contracts with merchants or millers for organic cereals. This should be done in advance of sowing, so that there is then a guaranteed market for that particular variety and quantity of wheat. Before this happens, it is necessary to build up mutual confidence. The merchant must be satisfied that the farmer is technically competent to produce quality organic wheat, and the farmer will want to know that the merchant will pay the agreed price on time.

Organic buyers of grain will have different requirements in terms of volume. It can be a small amount or vast quantities, on a one-off basis, or on a regular monthly supply. It is important that the organic producer has storage capacity and that the quality be maintained for several months.

As with other organic producers, organic crop producers need to know where their markets are, how to negotiate, and how to establish themselves as reliable suppliers through long-term relationships with buyers (Born & Sullivan, 2002). As demand for crops varies between the different types, farmers should speak regularly to their potential buyers to get feedback on the market and the demand. By establishing a relatively stable rotation of crops, the farmer will have time to plan his marketing.

Organic grains may be contracted prior to planting. Contracts vary considerably, some make producers responsible for grain cleaning or shipping charges or both, while others do not make the producer responsible for either.

Abattoirs, butchers

Just as the cereal farmer can sell his organic grain to the miller, so the livestock farmer can sell to the abattoir. The abattoir should be licensed by the Soil Association. The most important points to ensure are, first, that the organic animals are killed within a short time of arrival (that is, they are not kept in lairage overnight) and, second, that the chance of organic meat getting mixed up with conventional meat is eliminated. At one time, organic lamb was selling at 20 pence per kilogram dead carcass weight more than conventional meat, while all organic meat was in short supply and in demand. There must therefore be no loopholes whereby organic might be confused with conventional.

If the organic producer has a choice between a nearby abattoir that is not licensed but only entails a short journey, and a licensed abattoir that is more than six hours distant, then it is probably better to choose the closer abattoir, but it is imperative to ensure first that your animals will be killed quickly and also that there will be no possible mixing up of carcasses.

Finally, aim to get payment from the abattoir within two weeks. This is not unreasonable, and will help the all-important cash flow.

The first slaughterhouse in Denmark for organic meat was set up in 1994, for an organic pig farm nearby. It now produces organic cold meats and sausages. It advertises its products by informing the consumers that their products are made without additives, that all the meat is hand-cut, and that no mechanically recovered meat, blood plasma and extra meat proteins are ever used.

A Danish charcuterie company increased its sales of salami by 50% by being on the list of suppliers to a pizza chain that sells organic pizzas to British consumers (Organic Denmark, 2003).

Many small meat producers find that when they begin to direct-market their meat, it practically sells itself (Born, 2000). Word of mouth and minimal promotion enable them to sell quickly. The farmers need to realise that the market for this is relatively small and sales will soon dry up when all the 'easy' customers have been located and supplied. They would be well advised to hold back expanding production and investing in facilities until they have secured other sales channels.

Meat has a 10 to 14 day shelf life from the time of actual slaughter (not counting ageing time). Thus, locally grown and slaughtered meat will have a longer shelf life than that from conventional sources. This should be made into a selling point at the local butchers (Born, 2000).

Judging from the feedback of some butchers, a possible market for unpopular cuts of organic meat could be a 'premium' pet food manufacturer!

Co-operatives

For many farmers, co-ops can be a key to survival and success. In an undeveloped organic market, the farmer may not be able to take any advantage of economies of scale in processing, transportation and marketing. Furthermore, if there are only a few large buyers, the farmer may not have much power to negotiate a fair deal. To achieve a consistent presence in the market, the farmer will have to maintain a regular volume of quality product, and this may be difficult for the farmer acting alone. It is difficult for an organic farmer to run a farm operation and to devote time to develop skills in professional marketing or running a processing operation.

The traditional way for farmers is to form co-operatives. The advantages of these are that there will be a collaborative marketing effort and the group will be able to take advantage of size economies, maintain a steady flow of products, gain access to knowledge and professional expertise, increase bargaining power, and gain access to a broad, diverse customer base. This sytem has been tried, more for vegetables than for meat; and many such co-operatives have collapsed quickly (though more recent efforts are still under trial: see Chapter 10). Wolstenholme (1993) examines the reasons for this and draws three main conclusions:

(1) Lack of investment in the co-op's early stages, which often leads to important business and management posts being filled by the group's own members. They may well have neither the time nor the experience. An experienced business manager is therefore essential.

(2) Lack of diversity of outlets: for example, when the supermarket is the co-op's only customer then the group has been left with a lot of waste products which the supermarket has rejected because their specification demands product excellence and uniformity. It is therefore important to find other outlets for perfectly sound produce that has been graded out by the supermarket packer. Other markets include restaurants, shops, wholesalers and consumer groups.

(3) Within any group, some members put in more effort than others. The ones doing more resent the passengers, and the passengers feel taken over or ignored. Every member has to understand that the success of the co-op rests on each individual. Unless this is accepted by the individuals at the beginning, the co-op is likely to collapse under pressure.

Other problems of working together include lack of a common mission, commitment, and costs of group decision making.

The Organic Valley dairy co-op, based in Wisconsin (USA), has 633 farmers in 16 states and one Canadian province. They have 20 475 cows and 95 000 acres. They claim that their success stems from paying farmers a fair, sustainable and stable price. They sell via natural food stores and supermarkets. They have also been active in getting schools to offer organic milk (The New Farm, 2003).

Community supported agriculture (CSA), or subscription farming, is a direct marketing method started by organic farmers in the USA about a decade ago. Consumers subscribe to the harvest of a CSA farmer for the entire forthcoming season, and pay for their produce in advance. Consumers and farmers share the production and harvest risk of the farmer, and the bonus. There are variations to this scheme, which is a useful method of direct marketing.

Independent retail shops

This includes specialist shops, gourmet food shops, health food shops, natural food stores, and other farm shops, including delicatessens and food halls in large departmental stores.

The advantage of selling to local retail shops compared to having a farm shop is that the farmer does not have to be concerned with the day-to-day operations of running a shop and having to deal directly with customers. This leaves more time for farming. However, it is worthwhile spending time and effort on securing a good relationship with the shop owner, because it is vital that the organic product be prominently displayed, and that the information given to the customer about 'organic' be factually correct and interesting.

In order to supply organic food of a consistently high quality on a regular basis, it is necessary to have an overproduction on the farm. Only in this way can sufficient selection be made to ensure high quality, which in turn helps to maintain consumer loyalty. It also means that another outlet needs to be found for the seconds.

Before committing to an outlet, the farmer should be aware of the product lines and brands stocked by the retailer, the knowledge and efficiency of the sales staff, the presentation and product display, the clientele, the store's paying behaviour and integrity. In return, the farmer will need to accept that the store manager may attach specific conditions to items which are slow-moving, perishable or requiring considerable display space.

It is worth noticing what other products are likely to share the same aisle or shelf. In a study on how to increase purchases in the Minneapolis and St Paul markets in the USA, by including organic products, a mixed effect on sales was found in upmarket stores, but there was a significant positive effect on sales of dairy products, pasta, bread, cereal and carrots in a discount store (Thompson, 1998).

For gourmet food retail stores, it is important to be aware that there will be strong competition from other premium products being sold. The farmer should also be aware of the customer profile and (by simply looking in shopping baskets) get some idea of current preferences. Not many gourmet food shops or health food stores have fresh meat counters, whereas restaurants require regular supplies of fresh meat. Fresh and frozen meat are two different markets and require different distribution channels.

There is a significant relationship between attitudes (positive or negative) of store managers and observed quality of the produce (Sparling, *et al.*, 1993). If the store manager has a low opinion of organic produce, this will significantly affect sales. In-store handling was also a factor affecting the quality of produce. The stores that were more successful in selling organic produce mingled organics with conventional produce. In Denmark, pre-bagged produce was found to be unpopular. When a Danish chain of supermarkets used stickers to differentiate its organic produce from non-organic, and sold the items by weight, sales increased by 300% over two months (Organic Denmark, 2003).

Apart from when they are shopping in large supermarket chains, many consumers want to buy local products. Products which tell a story sell best: for example, 'local farmer' is a selling point, particularly for consumers from other regions (Roos, 2002). The farmer has to work with the retailer to market the farm, the products and the organic farmer's own local status. The farmer might wish to provide materials to the store, about the farm and farm produce. These could be photographs, a brochure or just a typed sheet of background information, and can be included in the retailer's promotional materials.

The leaflets, which should be uncomplicated, may include a 'box' for the name and address of any customer requiring further information about the farm, its products and outlets. Feedback on customers' preferences may also be obtained through these leaflets, and may be of use when production is being planned. Trial packs, in the case of certain products, may be a useful promotional tactic.

Large supermarket chains

Over the past few years, the UK supermarkets have been accused of being the main culprit in changing the face of the high street. The first generation of supermarkets, about 30 years ago, was relatively small and in town centres. As they expanded and cut their costs through self-service, bulk buying and heavy merchandising, they began to replace the small, traditional independent grocer. They expanded to out-of-town sites with free parking, and took the customers with them, thus threatening the health of the high street. Recently, they have tried new formats, introducing small stores back into towns, which carry ready meals, basic grocery goods and lunch-time snacks.

Compared to other European grocery retailers, UK supermarkets have the highest operating margins. Therefore, in the UK, large supermarket chains are dominant, and this is where most consumers buy their organic food. The various branches operate according to a policy laid down centrally, whereas some supermarket managers in Europe have a greater degree of flexibility, allowing them to purchase food locally. The same has been found to apply in Colorado, USA, where the handling of small quantities of slow-moving, perishable produce was difficult for central warehouses.

There is considerable debate within the organic sector as to the 'rightness' of selling to supermarkets. Most organic farmers would like their products to be sold in their local towns and not transported backwards and forwards across the country. They are also apprehensive about the power and ruthlessness of the supermarket buyers, who are not seen as being user-friendly. That said, the major multiples are the main sources for food purchases and their importance continues to rise. Therefore, if the sale of organic products is going to increase and reach a significant proportion of the population, it must be via the supermarket.

In the supermarket scenario, organic farmers must work hard to establish and differentiate their product from others. Many brands of organic product are now on the market, with marked differences in quality and price. Indeed, the organic farmer is now competing not only with conventional but also with other organic producers. For example, many premium non-organic products are sold on other merits: hand-picked, left to ripen naturally, outdoor reared, never frozen, no added water or colouring ... Similarly, organic farmers must ensure the highlighting of the 'organic credentials' of *their* produce: for example, free of pesticides, antibiotics and additives, GM free, gently stoneground, no growth hormones, no added protein, 'local', delivered fresh within 24 hours ...

Specific problems relating to meat

Producers deciding to target the large food retailers should consider the high volume, the quantity of meat and the marketing services that will need to be provided, such as prepackaging and delivery. In the UK the supermarket chains have formed partnerships with many organic farmers.

Hunt (1989) states that for Safeway or any other major multiple to launch and expand the sale of organic meat it requires guaranteed volume, guaranteed quality and acceptable price levels.

Guaranteed volume

The supermarket needs to offer organic meat at the same level every day of the year. Hunt (1989) mentioned that, as a supermarket buyer, he had been approached by several suppliers offering 'one tomorrow, one in two months' time and two for Christmas'. Safeway required ten cattle per week. To fit into Safeway's distribution system, these cattle would need to be processed through one abattoir, which should be Safeway and Soil Association approved. He believed it to be unlikely that one farmer could supply this number every week of the year. However, the example of Chisel Farm (see Chapter 10) has demonstrated otherwise. Hunt suggested that producer groups be set up in conjunction with a central slaughter point. This, however, would create the additional problem that the producers and the abattoir would have to be located so as to ensure that the transport of livestock was both practical and humane. This problem is now in the process of being solved (see FOLMG, Chapter 10): with what success remains to be seen.

Guaranteed quality
Supermarkets have a carcass specification for weight conformation and fat content, which ensures that the meat bought is commercially viable and provides the end product that the consumer requires. The supermarkets are able to insist on their own quality regulations for produce, because they are so dominant in the market place. Until such time as organic farmers can co-operate and become equally powerful, they will have to abide by the supermarket standards. At present, they seem to be under pressure to produce regularly shaped and sized, blemish-free products, on a mass scale, available all-year-round and often transported long distances. Despite these pressures, however, British organic farmers are not contract farmers to the extent that many conventional farmers are.

Acceptable price levels
This is of crucial importance. Jones (1989) observed that meat will always carry a lower premium than vegetables. Purchases of vegetables by customers are often made casually, with little attention being paid to the price per kilogram. This contrasts sharply with purchases of meat, where the most acceptable pack is sought carefully. Underlying this is the fact that most purchasers know the approximate price of the meat for a traditional Sunday lunch, but have far less idea of the cost of the vegetables which accompany it. The basic determinant of acceptability is price. This fact will permanently limit the premium available on organic meat. Hunt (1989) is in agreement with this. A very small sector are already buying organic meat at up to double the price for conventional meat, but these customers have incomes to match their principles. An acceptable level will not be twice the normal price.

By selling through the supermarket, rather than to an abattoir, several additional costs are incurred which increase the end price but not the price received by the farmer. These additional costs may be summarised as follows:

● Additional costs at the abattoir because the livestock and carcasses have to be segregated; these costs can be reduced when volume increases.

- Additional costs because retail packs of organic meat would be produced away from the supermarket, using environmentally friendly packaging.
- Additional costs because the multiple retailers require two hind-quarters for every fore-quarter – ideally a three legged beast! The spare fore-quarter will have to be sold in the conventional market at conventional prices, thus increasing the cost of the remaining 75%.

If the organic producer requires a 15% premium, then the additional costs listed above could increase the mark-up on organic meat to 50%, and the price structure would be unacceptable, except to a wealthy minority. This would be in conflict with the organic movement's aspirations; so, to prevent this, all additional costs must be kept to a minimum. If the UK organic licensing bodies were to reduce their levies or the number of stages at which a levy must be paid, a proportion of the additional cost might be lessened; for instance, at present, levies may be paid on beef at four stages – the farm, the abattoir, the packaging or processing plant and the retailer.

Supermarkets and other buyers insist that food described as 'organic' should be labelled with one of the logos approved by UKROFS (United Kingdom Register of Organic Food Standards). The term 'organic' now has legal status, and one of the reasons behind setting up UKROFS was to protect the consumer from misleading use of that term. A second reason was to bring the standards of the various organic bodies such as British Organic Farmers and Growers into line, again so that the public was not misled. It is also wholly reasonable that 'organic' farms should be inspected regularly to see that the regulations are being observed. In order to acquire an organic label and to keep it, farmers have to pay money. This was satisfactory to the farmer while the organic label commanded a premium, which was the case initially. Now, with the reduction in the organic premium, organic farmers feel that the cost of the label is swallowing up their profit margin.

Some sort of compromise needs to be reached. The buyer and the consumer must have the guarantee that what is called 'organic' really is. So either the organic associations such as the Soil Association have to reduce their charges or, conceivably, the government may be asked to make a grant to cover this aspect.

Supermarket chiefs (Love, 1989; Hunt, 1989) sound enthusiastic about increasing the volume of organic produce stocked and sold by the supermarkets, particularly as they feel that organic meat enhances the overall image of meat. If organic produce can also provide an improved flavour, this would of course be advantageous.

There can be a problem of competition between organic and conventional products. The question the customer asks is: why is the organic product more expensive than its conventional competitor? This question is best answered by informing the public as to what 'organic' stands for. Conventional products may be more profitable per kilogram, but supermarkets depend on new products, high quality lines and more choice, to get customers to come back. It is not necessary to frighten people off conventional products, nor to convince them that organic products taste better, in order to encourage them to purchase.

The answer has to be for supermarkets to explain the rigorous and appealing principles guiding the production of organic food, such as sustainability, less environmental contamination, improved animal welfare and healthier food. Many customers are sensitive to issues of welfare, residues and food safety. Indeed Germany would not have contemplated a ban (now lifted) on the importing of British beef because of fears about BSE (bovine spongiform encephalopathy, or 'mad cow disease') if the beef had been produced under the nutritional regulations required by the Soil Association.

Similarly, the widespread and devastating outbreak of foot-and-mouth disease in Britain in 2001 may have pushed consumers, farmers and government to a more serious consideration of the humane and ultimately sensible organic option. Under the Soil Association regulations, organic farmers are not allowed to buy or sell in markets. This would have prevented the widespread dispersal of sheep, particularly, throughout the country, which would have made local confinement of the disease much easier.

References

Born, H. (2000) *Alternative Meat Marketing*. ATTRA on http://www.attra.ncat.org

Born, H. & Sullivan, P. (2002) *Marketing organic grains*. website as above.

Brassington, F. & Pettit, S. (2003) *Principles of Marketing*, FT, Prentice Hall.

Cooke, M. (2002) Importing organic meat and dairy products and retail markets in the UK. Organic Meat and Dairy Products Symposium, FAO. on http://www.fao.org

Crothers, L. (2000) Australia, Organic market continues to expand. GAIN Report, USDA on http://www.ers.usda.gov

Dimitri, C. & Greene, C. (2002) *Recent growth patterns in the U.S. Organic Foods Market*. Economic Research Service, USDA.

Environment and Natural Resources, Series 4 (2002) Certified organic agriculture–situation and outlook, FAO, on http:www.fao.org

Fanatico, A. & Redhage, D. (2002) *Growing your range poultry business*: Entrepreneur's toolbox www.attra.ricat.org

Finke, J. (1989) Marketing from the Farm. The Case for Organic Agriculture, *Proceedings of the 1989 National Conference on Organic Agriculture*, 36–9.

Hunt, M. (1989) *Organic Meat Markets: a major food retailer's approach.* Organic Meat Production. Chalcombe Publications, Maidenhead.

Jones, E.O. (1989) The prospects for organic meat. *New Farmer and Grower*, 25, 30–31.

Krucsay, W. (1999) Austria, *Organic production and marketing of organic products.* GAIN report, USDA on http://www.ers.usda.gov

Love, J. (1989) The Supermarket Response. The Case for Organic Agriculture. *Proceedings of the 1989 National Conference on Organic Agriculture*, 40–42.

Marriage, M. & Starling, W. (1989) Cereal Quality, Storage and Marketing. The Case for Organic Agriculture. *Proceedings of the 1989 National Conference on Organic Agriculture*, 66–70.

Organic Denmark (2003) Danish salami on British pizzas; Unpackaged organics – a success; Focus on organics increases sales; Campaign success; Nomad campaign for organics in Danish retail trade, all on http://www.organic-denmark.com

Roos, D. (2002) *Marketing for independent retailers. Cooperative Extension Service*, North Carolina State University, on http://www.ces.ncsu.edu

Soil Association (2003) *Organic food and farming report.* Soil Association Bristol.

Sparling, E., Wilken, K. & McKenzie, J. (1993) *Marketing fresh organic produce in Colorado supermarkets*, on http://www.sarep.ucdavis.edu

Success with a Small Food Business (1994) Foodsense series, Food Standards Agency, Hayes, Middlesex.

The New Farm (2003) *Organic valley achieves record growth in 2003.* Locally grown beats organic in consumer survey, on http://www.newfarm.org

Thompson, G.D. (1998) Consumer demand for organic foods: what we know and what we need to know. *American Journal of Agricultural Economics*, 80, No 5.

Wier, M. & Calverley, C. (2002) Market potential for organic foods in europe. *British Food Journal*, 104, 45–62.

Wolstenholme, J. (1993) Organic producer co-operatives. *New Farmer and Grower*, 39/40 20–22.

Further reading

Hingston, P. (1988) *The Greatest Little Business Book*, Hingston Associates, Perthshire, Scotland.

10 Progress by Organic Farmers

All farms are different, but certain problems recur, some to do with husbandry and some to do with marketing. In this chapter, problems that are common to several farms are described, in the belief that they will be of interest to farmers treading the same path to full organic production. Actual organic farms are described and the progress that has been made towards a solution is charted.

Insufficient organic grassland during conversion

The problem occurred on a large 1300 ha farm on chalk and clay in Wiltshire during the conversion process. The stock on the farm consisted of 1550 ewes and their lambs and 75 Hereford × Friesian suckler cows and their calves. The available symbol grassland was 200 ha of chalk downland permanent pasture and 30 ha of grass/white clover leys in conversion. The remaining grassland was part of the arable rotation and did not qualify at this stage as organic according to the symbol scheme. If the sheep and suckler cow enterprise were to qualify as organic then to meet their nutritional requirements another 70 ha of organic grassland were needed; or the stock numbers could be reduced, but this would reduce the profit.

The dilemma was resolved by making the farm part of the environmentally sensitive area (ESA) scheme and also sowing an extra 280 ha of grassland. As his farm was now part of an ESA, the farmer was paid to reduce his stock numbers so that they then had sufficient organic grassland. This meant that although fewer lambs and beef cattle were sold per ha, the grant from the ESA scheme more than compensated for the reduction in output.

The value and methods of introducing white clover

White clover is the pivotal legume in organic grassland. It fixes nitrogen from the atmosphere which it will then release into the soil when its roots decay; this released nitrogen substitutes for the fast-acting nitrogen supplied by artificial fertiliser such as Nitro-chalk and nourishes the companion grass plants. Estimates of the quantity of nitrogen fixed by white clover vary between 200 and 300 kg per ha per annum, which is a considerable amount. However, what is less certain is actually when this nitrogen is released into the soil to become available.

The second benefit of white clover is that it increases animal production, whether in terms of higher milk yield for dairy cows or faster growth rate in beef cattle and lambs. The extra production is derived from the higher intake of white clover compared with grass of the same digestibility and also from its higher protein content.

There are two basic methods of introducing white clover into the sward. One is to plough the field up and sow it as part of a seeds mixture, and the second is to slot seed into an existing grass sward. In both cases the clover proportion can be changed by subsequent management.

On a conventional farm, the recommended technique for slot seeding is to check the grass growth with a weak solution of paraquat and then to sow the white clover in slots. If germination of the clover is favoured by sufficient moisture, then the clover seedlings can establish and start to grow away before the grass recovers from its dose of chemicals. This option is not open to the organic farmer. One organic farmer, farming on Cotswold brash which has a naturally high pH, decided to slot seed white clover into grass without using paraquat. The sward was grazed down to 3 cm in August 1990 and then white clover was seeded into slots cut into the sward. Since then the field has only been grazed by cattle. The proportion of white clover present, using a point quadrat to determine first hits, is shown in Fig. 10.1. More than 30% of the sward was white clover in March 1992 and in the following year 48% of the sward was clover as early as February. The amount of clover present was equally impressive in October, although white clover needs a warmer soil temperature than grass in order to grow actively.

The second and third instances of clover increase on organic farms concern a hill farm in Wales and another in Scotland. In both cases the farmers felt that they should plough up their fields and resow to a grass and white clover mixture, mainly because the swards were thin, low yielding and only had small quantities of white clover present. In the case of the Welsh hill farm which was over 400 m high near

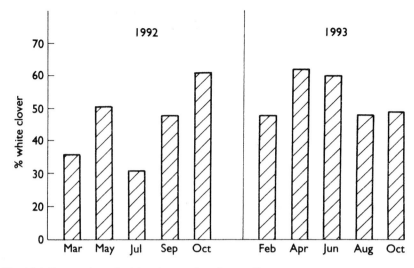

Fig. 10.1 Proportion of white clover after slot seeding.

Plynlimmon, the only animals available for grazing were sheep. However, before deciding to plough up and resow, the Welsh farmer felt that two improvements in management technique were worth trying. One was to increase the soil nutrient status using farmyard manure and the other was to revert to rotational grazing rather than set-stocking with the sheep. Too often the symptoms of low soil nutrient status – a poor quality low yielding grass sward – are tackled without considering the cause. There was little doubt that sowing ryegrass and white clover without increasing the low levels of phosphate and potassium would have been an expensive error, with a similar poor quality sward as the end result.

Sheep, unlike cattle, select white clover and the consequence of leaving them on the same field, month in month out, is to decrease the proportion of white clover because it gets defoliated more frequently than the grass and thus has no opportunity to spread. By rotating the sheep round the fields, aiming to defoliate the whole field with the whole flock in a week, and then to rest it from grazing for 4–5 weeks depending on the season, the clover is given the same chance as the grass. The third important factor in promoting white clover in permanent pasture is to ensure that the sward is grazed down to about 3 cm in October. This enables light to reach the clover stolons, which encourages axillary growth and the spread of clover in the spring.

In 1991, at the start the proportion of white clover on the field was less than 5%. The figures for white clover in 1992 after the addition of farmyard manure and rotational grazing were 6% in April, 39% in June, 50% in August, dropping to 19% in October.

The Scottish less favoured area farm near Kinross had similar poor quality swards on the inbye land. The solution was the same, to use plenty of farmyard manure to improve the soil nutrient status, to alternate sheep grazing with cattle grazing annually, and to ensure that the grass was not allowed to shade out the clover during the long winter months. The result has been similarly successful: more clover, a higher yield and much less yellow-looking grass.

The control of internal parasites by grazing management

A tricky problem for organic sheep farmers to overcome is the control of internal parasites in their lambs, particularly if sheep are the predominant enterprise. Beef and dairy cattle suffer far less frequently from parasitic infections. Conventional anthelmintics can only be used if there is a demonstrable problem with parasites, by which time it is often too late to redress the growth rate of the lambs. Furthermore, even in conventional sheep flocks there is increasing evidence of growing resistance of parasites to the current range of chemicals. So far homoeopathy and herbal remedies, such as garlic or wormwood, have not provided a convincing cure for *Nematodirus*, *Ostertagia*, *Trichostrongylus* or *Haemonchus*. The best approach therefore is one of prevention, or at least minimising the challenge to susceptible lambs until they have had time to build up their resistance to internal parasites.

Parasitic eggs overwinter quite successfully in Britain, so that in order to ensure a clean pasture the field needs to be rested from sheep for a whole year. Fortunately, parasites that damage sheep do not affect cattle and vice versa, with the exception of *Nematodirus* which affects sheep and calves up to the age of six months. The easiest method of carrying out a policy of clean grazing is to have a rotation of arable, followed by sheep, followed by beef (Fig. 10.2a). This is fine if the land can all be ploughed up and used either for cereals or grass. It also helps if the number of livestock units for sheep is the same as for cattle. In most calculations seven ewes with lambs are equal to one cattle beast. This is based on animals eating in proportion to their weight. Seven 70 kg halfbred ewes weigh and eat the same as one 490 kg South Devon. If the ewes are very small, for instance Welsh Mountain ewes can average 30 kg liveweight, then account has to be

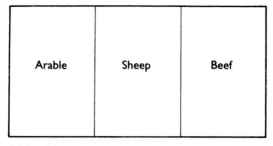

a) Mixed farm

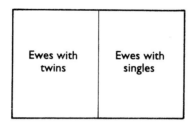

b) All sheep farm

Fig. 10.2 Clean grazing methods.

taken of this in calculating the eating power of the flock. The advantage of having the beef herd and sheep flock of similar eating power is that the area of grass devoted to each will be interchangeable between years. A hundred suckler cows can be balanced by 700 ewes.

One organic farm, in the Scottish Border region, of nearly 500 ha followed this clean grazing method scrupulously, with the result that the lambs fetched the best prices consistently, in competition with the conventional lambs, on the electronic system of marketing, a system where lambs can be sold before they leave the farm. The only deviation from the two-year rest from sheep comes on the permanent pasture land, which alternates between beef and sheep. A similar clean grazing policy is operated on an organic dairy farm in Dorset. Here the enterprises are dairy cows, cereals and sheep, and again the sheep will graze a field only every third year, with consequent benefit to the level of parasite infection in the lambs.

The upland organic farm in the hills of Wales was cited earlier in connection with white clover. Here the problem of clean grazing is more intractable, first because there are no cattle and second because none of the land is suited to cereals. The best approach, suggested in books, is to graze the ewes with singles separately from the ewes with twins each year (Fig. 10.2b). The theory here is that the single lambs fatten mainly on their mother's milk and therefore eat smaller quantities of infected grass than the twin lambs. The ewes with singles act like a beef herd. The weakness of this solution is, first, that the lactating ewes with singles still deposit parasite eggs on the pasture and, second, that the proportion of ewes with singles and twins varies each year and is unlikely to be half the flock anyhow.

A better solution is to replace sheep by beef cattle, money permitting, and gradually work towards similar flock and herd size in terms of eating power. This is what is happening on the Welsh farm, with the benefit that the lambs are now growing faster and being sent to market earlier, with a consequent reduction in stocking rate in the difficult months for grass of August and September.

Further results of introducing cattle are, first, to increase the clover content of the sward, because cattle prefer grass and, second, to increase the proportion of land required for hay or silage for the winter. Cattle often have to be fed hay/silage for six months of the year, whereas sheep require winter feeding for only two months. However, the cattle will produce much more farmyard manure, which can be put back on the pastures that were cut for hay or silage.

The phosphate story

During the course of an experiment to measure herbage production on organic farms in England, Scotland and Wales, it became clear that the most reliable predictor of herbage yield was the level of available phosphate in the soil (Newton, 1993). This relationship between herbage yield and available P held good over a range of sites: sites with varying numbers of grass growing days (based on altitude, summer rainfall and latitude and longitude); sites based on very different soils (clay, loam, sand and chalk); and sites grazed by different classes of stock (dairy cows, beef cattle, sheep and poultry).

Conventional advice maintains that there is no benefit in adding P to soils that contain available P of index 3 (26–45 mg/kg P in the soil). Beyond this level, there is unlikely to be an increase in herbage yield (Lockhart & Wiseman, 1993). However, the experiment cited above (Newton, 1993) showed a straight-line response up to levels of available P of 103 mg/kg (index 6). The equation derived from this work suggested that for every increase of 10 mg/kg available P in the soil, the yield of herbage dry matter will increase by one tonne. Work from Rothamsted (Cooke, 1987) had shown a similar response in yield for spring barley, potatoes and sugar beet up to levels of 160 mg per kg of available P in the soil.

One of the problems is that the phosphate fertiliser allowed by the organic regulations is rock phosphate, and research in Austria has shown (Lindenthal, *et al.*, 2000) that the application of rock phosphate had virtually no effect on subsequent crop yields.

Whereas any N and K that is applied to fields is immediately available to the growing plant and then tends to be washed down out of reach of the plant roots, a large proportion of the P that is applied is rendered unavailable to the plant by being bound into aluminium or iron compounds if the pH of the soil is low, and by calcium compounds if the pH of the soil is high. Although it may seem rather an expensive waste of money, at the time, that as much as 40% of the P applied in farmyard manure should be unavailable to the crop, it does stay in the soil and will become available to the crop in time.

This is well illustrated by two organic fields in Leicestershire (Newton, 1993). These were fields in the famous fattening pastures of Leicestershire and were shown to sustain a high annual stocking rate of beef cattle and sheep (three livestock units per ha), despite not having had any fertiliser applied in living memory, not even farmyard manure. And yet levels of available P and K were still index 6 and 7 respectively. It so happened that one of these fields had been the subject of an experiment in 1948 (Davies & Williams, 1948) and had been shown then to have a high level of available P, and had been cited by Harper (1970) 'as probably the most productive permanent pasture in the fattening districts of the midlands'.

The key to its continued high level of productivity over the recorded 50 years, despite having no artificial fertiliser or farmyard manure, was the high level of P and K and the higher N as well, because it was old permanent pasture. The high level of soil nutrients had been maintained, first, because it had not been cut for hay or silage in the last 50 years and so had no nutrients removed and, second, the high stocking rate of grazing animals had effectively recycled plenty of N, P and K.

The beneficial effect of available P is further illustrated by two adjacent fields on an organic hill sheep farm in Wales and by two fields on an organic Shropshire dairy farm (Fig. 10.3). Fields A and B on the Welsh hill farm were both clay loams, and field A (with 27.2 mg/kg available P) yielded 13.4 t dry matter per ha per annum, whereas field

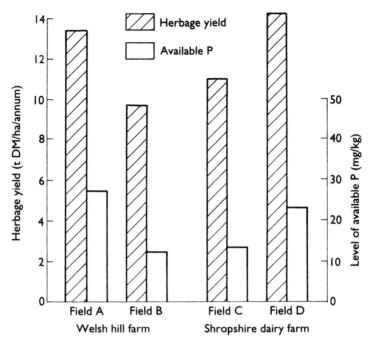

Fig. 10.3 The effect of available P on herbage yield (Newton, 1992).

B (with 12.2 mg/kg available P) yielded 9.7 t dry matter. On the Shropshire dairy farm, the fields were both analysed as sandy silts, and field D with the higher level of available P yielded 14.1 t dry matter per ha per annum, whereas field C yielded 11 t dry matter.

Some fields obviously have to be cut for hay or silage, and essential nutrients, N, P and K will be removed. Organic farmers cannot replace these lost nutrients by artificial fertilisers, but they can use either farmyard (mainly cattle) or poultry manure, provided the poultry manure comes from a poultry system that is free range and not a battery system. Figures for available P for an organic field in Wiltshire that was cut for silage in June 1993 and which then had poultry manure applied in spring 1994 show how effective this can be. In May 1993 the level of available P was 63.2 mg/kg, and by May 1994 the level had risen to 91.9 mg/kg. The nutrient value of different forms of farmyard manure and the causes of nutrient loss are well illustrated in a new DEFRA booklet (Making Better Use of Livestock Manures on Grassland, 2001).

Veterinary problems on organic farms

A great number of people have joined the organic movement because they feel that the welfare of farm animals has suffered in the frantic demand for more profit from intensive systems, particularly with pigs and poultry. At the same time, most people in the organic movement have a lively interest in alternative medicine, using both herbal and homoeopathic remedies where appropriate. Certain members of the veterinary profession regard organic farmers as foolish because they will not take advantage of modern veterinary drugs for their animals, and feel that such stock are a health hazard to neighbouring livestock on conventional farms.

It is stated quite clearly in the Soil Association Standards (1992) that:

'Animals should be sustained in good health by effective management practices, including high standards for animal welfare, appropriate diets and good stockmanship rather than relying on veterinary medication ... Treatment must never be withheld if an animal is seriously ill, suffering or considered by a veterinary surgeon or stockman unlikely to recover fully without treatment. Should treatment be withheld in these circumstances, the Certification Committee reserves the right to withdraw the Registration from that enterprise.'

So there is no question of organic animals suffering, without proper treatment. The problem for the organic farmer is that stringent experiments testing herbal and particularly homoeopathic remedies have not been carried out. Evidence of their efficacy comes therefore from farm treatment with no controls.

Lowman (1989) reports a trial in which organic cattle showed symptoms of copper deficiency, which was confirmed by blood samples. A homoeopathic preparation, *Cuprium metallicum*, was then sprayed twice weekly onto their organic barley. Two months later, blood copper levels had improved significantly, even though on analysis no trace of copper could be detected in *Cuprium metallicum*. In the same trial, organic calves were dosed with a homoeopathic remedy, *Teucrium marum*, against stomach worms, and no trace of worm infection was found, even though the calves were grazed on organic pastures which had been grazed by cattle for the previous five years and hence would be expected to suffer from heavy worm infections. Lowman (1989) stresses that whilst the use of homoeopathic preparations appears to have raised copper levels and to have prevented worm infections, no comparisons with conventional medicines were carried out nor were the potential problems quantified using control animals.

Halliday (1990), a Senior Veterinary Investigation Officer, records that he had an apparently effective response to the treatment of respiratory disease in calves, with a combination of *Bryonia alba* and *Hepar sulphuris*. In subsequent years, some animals with similar respiratory signs were left untreated as controls. They also recovered, apparently as rapidly and as effectively as the homoeopathically treated calves.

Day (1991) cites the case of mastitis in cattle where nosodes, used in preventive programmes, have reduced the incidence of mastitis and have made those cases which do occur easier to clear. Caulophyllum has been used to treat dystokia (calving problems) with success. However, Day (1993) is equally clear as to when a homoeopathic nosode is not advisable. He states that nosodes do exist for the prevention of clostridial diseases in sheep, but they have not been proved to be effective. Some people have used them with apparent success, but there was no evidence as to a significant challenge.

Selling organic meat

In the first half of the 1990s, at least four co-operatives were set up to sell organic meat, and each failed, often quite rapidly. Weisselberg (1990) described Welsh Organic Livestock Farmers (WOLF), which was set up because 'there is a desperate need for an infrastructure to co-ordinate the fast developing market for organic meat'. He went on to report that 'over the next 12 months the co-op expects a throughput of 10 000 lambs, 4000 store lambs, 500–600 head of beef cattle and 200 store cattle'. Projection for the future suggested that WOLF could possibly quadruple supply within the next year, and that by the year 2000 10% of the country's meat demand would be met by the organic sector. About 70% of WOLF's sales were to the large wholesalers and abattoirs, the other 30% were to minor wholesalers. At that time (1990), the premium over conventional prices for the organic producer was 15–25% on lambs and about 25% for cattle.

WOLF, after a lifespan of no more than a year, was followed by another co-operative based in Wales; there was then 'Honest to

Goodness Food' and a small meat co-operative based in Dorset. After an initial period of enthusiasm and confidence, these co-operatives ceased trading.

Various reasons have been put forward for failure: absence of a full-time professional manager, lack of a reliable supply of quality organic meat and no lasting worthwhile sales of organic meat, either by the large multiples or by the high street butcher. The supermarkets mentioned that the insistence by the Soil Association on a licence for the abattoir and then another one for the meat-packing depot meant either a reduction in the supermarket's profit or a hefty increase in the price of organic meat to the customer. The message was clear: the organic movement was pricing itself out of the market.

Logically and logistically, the obvious way for a relatively small number of geographically scattered organic meat producers to sell meat to supermarkets on a regular long-term basis is to band together.

In 2002 key organic livestock producer groups in the UK set up a livestock marketing federation, FOLMG (the Federation of Organic Livestock Marketing Groups), with four objectives:

- Expansion in the number of outlets selling organic meat.
- Increased volume and quantity of supplies.
- Better co-ordination of supplies to meet the needs of the market and improved market intelligence.
- Collective work to achieve viable farm-gate prices.

(Soil Association, 2003)

Selling organic milk

The Farmers' Dairy Company (FDC) has had a longer existence than any of the organic meat co-operatives. It has now been trading for nearly four years. FDC appointed a professional manager, and the advice was to sell value added products that did not compete with each other. For instance, one dairy farmer bought a milk packaging machine to sell organic milk, one sold milk to a local creamery to make organic cheese, one sold organic yoghurt and one intended to sell organic cottage cheese. Again, FDC has been subjected to the vagaries of the supermarket buyers. It was probably detrimental to its success that one of the most successful organic firms, Rachel's Dairy in Wales, did not become a member of FDC. As Dowding (1992) argues:

'There is still a large volume of unfulfilled customer demand ... we do not serve any purpose by unnecessarily competing amongst ourselves to try to fulfil it. A structured approach is needed.'

Marketing organic milk in some European countries has also had its problems (Dunn, 1992). Bioland in Germany, the largest organic farming organisation in the country, with 2200 members, projected that the quantity of organic milk produced would reach 35 million litres, but that only half of this would find a market as organic milk and receive a premium price. Danish Nature Milk are signing no more contracts with organic dairy farmers, following an initial lucrative five year contract offering 35% premium to organic producers, because they feel that the market for organic dairy products in Denmark is not expanding. Money is being spent on advertising and promotions in order to increase public awareness of organic dairy products in Denmark, and the capital for this campaign is being provided by a levy on marketed organic milk. Organic milk in supermarkets has to compete with conventional milk which is heavily discounted. The crucial problem for organic producers in the rest of Europe, as well as in Britain, is the price of organic products in the shops. The price must be sufficient to allow organic farmers to stay in business, but not so high that the customer demand is extinguished.

Yeo Valley Organic Company

The organic dairy industry has been greatly boosted by the success of the Yeo Valley Organic Company. Since its foundation, the number of dairy farmers in the Organic Milk Suppliers Co-operative (OMSCo) increased from five in 1994 to over 200 in the year 2000. One of the reasons for this increase has been a three year rolling contract with a price guarantee for organic milk, which was based on a calculation of the true costs to an organic farmer of producing a litre of organic milk. Although the price per litre that the organic dairy farmer receives is currently higher than the one his conventional neighbour receives, the margin will fluctuate depending on the price for conventional milk.

Although the Yeo Valley Organic Company sells products other than yoghurt, such as organic milk, butter, cream, crème fraîche and fruit compotes, it is the success of organic yoghurt that has caught the public's imagination.

In two years, organic yoghurt sales rose from £17.2 million to £39.1

million, and this is now 6.6% of total yoghurt sales in the multiple grocers. Yeo Valley organic yoghurt sales have risen from £10 million to £22.6 million in two years (1998–2000), and this is 57% of the total sales of organic yoghurt. Yeo Valley Organic produces the top selling natural product and the top selling wholemilk product, with Yeo Valley Organic Strawberry.

The company firmly believes that the organic market will increase and that, now organic food is no longer just a niche market, the large food companies and retailers will start to sell their own organic brands. This will force prices down and so encourage farmers and processors to take short cuts, which will put the integrity of the organic industry at risk. The strongly held conviction at Yeo Valley is that the price of organic food to the consumer does not have to be higher than 20% more than its conventional counterpart, a target which can be achieved by an efficient company. However, organic products will continue to be slightly more expensive, because the raw materials are more costly, owing to organic farming techniques, reduced yields and decreased animal stocking rates. Furthermore, the farmers should be guaranteed a reasonable profit margin.

Organic cheese

Organic cheese is now being produced in increasing quantities to satisfy demand. Gary Jungheim of Country Cheeses, Tavistock, has seen a 10% increase in 18 months.

Organic cheese is made to widely varying recipes, from triple cream soft cheese to hard full-flavoured sheep's cheese. There is also organic Double Gloucester, a blue Jersey milk cheese, an organic blue Stilton and rinded and non-rinded Cheddar. Real farm cheeses vary considerably in taste according to season and where the animals are pastured. Organic adds other nuances to a variable product. Because organic milk attracts a premium price, this is reflected in the selling price of organic cheese, which may be up to 30% more expensive.

Chisel Farm, Dorset

Background
In 1988 the organic unit on Chisel Farm measured 120 ha, from which beef cattle and lambs were produced. Twelve years later (2000) the

area devoted to grassland, both temporary and permanent, was 571 ha and the arable area was 120 ha. The build-up in the size of the unit was gradual and another 100 ha were added in 2001, with most of the unit being rented under the Farm Business tenancy, plus some land being rented from the RSPB (Royal Society for the Protection of Birds), the National Trust and English Nature.

On this large organic unit there was a dairy herd of 300 cows, mostly Shorthorns, 310 beef animals of one year or more, 178 beef animals of less than a year, 425 breeding ewes (mainly Mules) and 1000 to 1500 store lambs. The number of store lambs bought in depended on grass availability.

Productivity

From this large unit three organic beef cattle were slaughtered each week, except for the six week period before Christmas, when eight per week were sent to the nearby abattoir (6 miles). Thirty organic lambs were slaughtered per week from the beginning of September to the beginning of July.

The carcass weight of the heifer averaged 240 kg to 250 kg by 16 to 20 months and the steers were slaughtered by 25 months with a carcass weight of 280 kg to 320 kg. The average carcass weight of the lambs was 18 kg to 20 kg, and these graded at 3L, 3H and 4L. The Blackface lambs averaged 15.7 dcw.

Breeds

For the cattle, native breeds were preferred, with Shorthorn, Angus, Devon, Hereford and Welsh Black predominating. The sheep were a mixture, but Texel and Charolais were the most popular terminal sires.

Feeding policy and management

The yearling cattle were out at grass during most of the year, except for the winter months when they were fed only silage. The older cattle received silage during their second winter, with 500 g of wheat daily, and were finished before the spring.

The sheep were never housed, nor were they fed anything other than grass. As for grazing, they were either pastured on the hilly fields or followed the cattle.

Health problems

The main problem with rearing the beef calves was pneumonia. To try to prevent this, a specialist calf unit was constructed, so that the calves never came into contact with older cattle. Scouring was not a problem.

Parasite control in the sheep was achieved by a clean grazing system. Store lambs that were bought in from outside were invariably quarantined.

Business management

Verbal agreements with retailers for the supply of organic beef or lambs were made a year in advance. This allowed time for the management and growth of the necessary stock. Stock were sold to the retailer directly from the nearby abattoir. This meant there were no delivery or collection charges for Chisel Farm. The then manager, Karl Barton, maintained that verbal agreements were more satisfactory than written contracts, which often contain hidden penalties.

Suitable store cattle and lambs were bought by Chisel Farm through Alder King, and any stock that were not regarded as of a sufficiently high standard were sold at conventional prices as non-organic meat through Alder King.

The secret of success

Karl Barton managed to do what the industry had urged the organic sector to do for a long time: to supply the market with high class organic beef and lamb on a regular basis. At Chisel Farm it was no longer a question of 'here are six beef animals this week, there probably won't be any more for several months', but a regular supply every week of the year. In order to achieve this, Karl Barton had an enterprise that was large enough to maintain a constant supply. He applied a high level of stockmanship and devotion to detail, and insisted that any stock below standard were not sold as organic. The greater proportion of his meat was sold as organic at prices that were better than those currently available for conventional animals.

Good Herdsmen

This organisation, run by Josef Finke in Ireland, has developed into a large-scale organic business, in a similar manner to that of Karl Barton at Chisel Farm in Dorset.

Good Herdsmen, which co-ordinates the selling of beef and lamb from 45 organic farms in Ireland, now sells to 60 Tesco supermarkets on a weekly basis. There is a continuous lamb supply from May to

February, and beef is sold de-boned and vacuum packed and cut to the customer's specifications, all certified and labelled with the Organic Trust symbol.

Josef Finke is described as having done more to raise awareness, develop markets and preserve standards of organic food in Ireland than anyone else (An Bord Bia – Something Special, 1998). He has built his own mill, launched his own range of organic biscuits, woven his own organic cloth and blankets, and sold organic oats to Nestlé and organic wool to Esprit. He has secured contracts for the sale of 12 500 organic lambs to Frankfurt, 10 t of organic wool to Esprit, and is negotiating the sale of 600 sweaters to Greenpeace.

Finke, from Ballybrado House in Tipperary, has also begun to delegate to other companies, with a bakery in Bray, County Wicklow, and knitwear produced by Kerry Woollen Mills.

The Watermill, Little Salkeld, Penrith

In October 1974, the Watermill, Little Salkeld (in the Eden Valley), together with its traditional machinery, was taken over by Ana and Nick Jones. They had always been interested in the preservation of old crafts and skills and were determined that the mill would be kept, made to work and not be converted into stables, a marina or a craft shop. Since neither had previous experience of flour milling, Nick Jones embarked on a ten day training stint in Wales, and a millwright from Staffordshire was called in to get grinding underway.

A grant was obtained from the English Tourist Board in order to get the mill back into production and also to convert the milking shed into a tea room: this led to a commitment to open to the public three afternoons a week, for ten years. In 1975, with the tea room completed, as well as a car park for about 20 cars and extensive publicity from the Tourist Board, the Watermill opened fully to visitors and rapidly became a popular venue.

The Joneses then set about looking for customers for their flour, and were soon delivering to Yorkshire and Manchester, as well as Cumbria. After several very busy years involving milling, delivering, baking, and running the tea room, they were able to pay off all their loans. However, efforts to hand over the milling, an increasingly demanding job, proved difficult, since millers with the necessary skills and stamina were not readily available for employment.

In 1985, with a young family and burdened by a heavy workload, the Joneses closed the tea room. Unfortunately, this coincided with a recession and an accompanying drop in demand for organic products. After a difficult five years, it was decided to re-open the tea room – independently of the Tourist Board this time – as well as an organic shop in the tea room. Within the shop it is now possible to buy a large range of dried organic goods: dried fruit, pulses, grains, oats, nuts and seeds, plus ingredients for cooking and baking. Ana Jones also makes jerseys, hats, gloves, cushions and scarves from the wool of her Suffolk sheep. The wool is commercially spun and then dyed and knitted or woven.

Flour

The flour milled at the Watermill is made from English wheat. It has a good taste and makes excellent bread. It is lower in gluten than imported wheats and is therefore not so demanding on the digestive system. At the time of writing, Canadian or European wheat is cheaper to buy than English wheat, but there is the added energy cost of transporting it long distances.

Wheat can be scientifically tested for protein level (gluten content), bushel weight, size of grain and Hagberg number, but this does not indicate the colour of the wheat or if it will make into wholesome bread. To determine whether the wheat sample will produce good bread, a small sample is ground in a coffee grinder, at the Watermill, and made into a loaf. If this proves satisfactory, then a sack of wheat is put through the mill and the loaves tested again by eating them. Only then are 20 t bought. Of the wheat milled at the Watermill, more than half is biodynamic and comes from two farms. The ideal solution would be for the miller to talk to the farmers directly, rather than go through the middleman (the grain merchant), and then to be able to stamp each bag of flour with the name of the farm where the wheat was grown. Most of the wheats milled at the Watermill are winter wheats.

Type of mill

The watermill is powered by two overshot wheels; the bigger wheel does the milling and the smaller wheel is used to clean wheat and to extract white flour. The millstones come from France and are the best millstones available. Stone milling is gentle on the wheat and the wheat germ gets ground into the flour. The end result is a flour with firmness to it and with a good texture.

Types of flour produced

At the Watermill, stone milling produces different types of flour: 100% wholewheat flour, 85% wholewheat flour which has had the bran removed, and white flour which consists of 50% of the wholewheat flour, which has been passed over a sieve. The remaining 50% is middlings, semolina and bran. There is a Granerius mix which has malted wheat added (malted wheat is wheat that has been sprouted and then dried). There is a Special Blend, which is wholewheat flour plus sunflower and sesame seeds and soya flour. This is for vegetarians, as it has a higher protein content. Then there is a Four Grain Blend, which is a mixture of oats, rye, barley and wheat flakes, and Millers Magic which is 50% wheat and 50% rye and contains less gluten. There is rye and barley flour and finally there is self-raising flour, which has bicarbonate of soda added to wholewheat flour.

At the Watermill, approximately 120 t of wheat are ground each year, plus 20 t of rye and 5 t of barley for people who cannot eat wheat.

References

Cooke, G.W. (1987) *Fertilizing for Maximum Yield*, 3rd edn. Blackwell Science Ltd, Oxford.

Davies, W. & Williams, T.E. (1948) Animal production from leys and permanent grass. *Journal of the Royal Agricultural Society of England*, **109**, 148–65.

Day, C. (1991) An Introduction to Homoeopathy for Cattle. *New Farmer and Grower*, **31**, 21–3.

Day, C. (1993) Homoeopathy. *New Farmer and Grower*, **39**, 14–15.

DEFRA (2001) Making Better Use of Livestock Manures on Grassland.

Dowding, O. (1992) BOMP [British Organic Milk Producers] and the Organic Market Place. *New Farmer and Grower*, **37**, 28.

Dunn, N. (1992) Boom or Bust. *New Farmer and Grower*, **34**, 13.

Halliday, G. (1990) Animal Health and the Organic Livestock Producer. *New Farmer and Grower*, **28**, 14–16.

Harper, J.L. (1970) Grazing, fertilisers and pesticides in the management of grasslands. In: *The Scientific Management of Animal and Plant Communities for Conservation* (eds E. Duffey & A.S. Watt). Blackwell Science Ltd, Oxford.

Lindenthal, T., Spiegel, H. and Freyer, B. (2000) Effects of long-term p-fertiliser application with different p-types and p-rates on p-balances, soil p-contents and yields. *Proceedings of the 13th International IFOAM Scientific Conference*, Basel, Switzerland, p. 23.

Lockhart, J.A.R & Wiseman, A.J.L. (1993) *Crop Husbandry and Grassland*. Pergamon Press, Oxford.

Lowman, B.G. (1989) *Organic beef production. Organic Meat Production in the 90s*. Chalcombe Publications, Maidenhead.

Newton, J.E. (1992) *Herbage production from organic farms*. Research Note 8, Elm Farm Research Centre, Newbury.

Newton, J.E. (1993) *Herbage production from organic farms. Report of second year of experiment*. Elm Farm Research Centre, Newbury.

Soil Association Standards (1992) Revision 5, Soil Association, Bristol.

Soil Association (2003) *Organic Food and Farming Report*, 81 pp.

Weisselberg, T. (1990) WOLF stalks the market. *New Farmer and Grower*, **28**, 18–19.

11   Organic Farming – the Future

Will Best

When Jon Newton wrote the first edition of this book in 1995 he asked me, as an organic farmer of ten years' experience, to contribute a final chapter giving my thoughts on the future of organic farming. At that time the farming was carried out by a small band of enthusiasts, most of whom were also engaged in some form of processing and/or marketing of their products. Both consumers and producers tended to be seriously 'green', and were regarded by the rest of the community as rather odd. We believed (as I hope we still do) that our way of farming was the right one, and wished to see its widespread adoption. The question was: to what extent was this likely to happen? My somewhat pessimistic conclusion at the time was: not much. I just felt that mainstream consumers and governments were not likely to take on board the environmental message and create enough demand and support to encourage hitherto mainstream farmers to convert. I thought that the only way for organic farming to become seriously noticed would be for the standard of the farming to become so high that lots of farmers would see the value of it and want to change.

Well, it looks now as if I was wrong. In these few years there has been an extraordinary shift in attitudes, as much within the farming community as outside. The figures speak for themselves: by 2002 there were one million acres of organic land in the UK, which is ten times as much as in 1997, while retail sales of organic food grew from £200m in 1996/1997 to over £600m in 1999/2000. Supermarkets are now expected to stock a whole range of organic foods and drinks, celebrity chefs all endorse organic, and organic farmers are now regarded by most of their peers no longer as oddballs but as rather sensible, astute even. (This takes a bit of getting used to.) Interestingly, you no longer have to be an all-round committed 'Green' either to grow or to consume organic food. (This is good news for makers of large 4×4 vehicles.)

I don't think that I was the only person to be taken by surprise by this explosive growth. So, how did it happen? Clearly, several factors worked together to drive it. The biggest was BSE. This, Britain's most expensive ever peacetime disaster, has been catastrophic for much of the country's livestock farming. And, although many recently converted organic farms were afflicted by the disease, the message which came over loud and clear was: if all farming had been to the Soil Association standards, BSE could never have happened. This, combined with many other food scares, mainly connected with agrochemicals and animal disease, and recently the GM issue, caused a major shift in consumer perceptions of food and caused many thousands to see organic food as imperative for themselves and their families.

Initially, nearly all the demand was met by imports, which stirred government to make funds and free advice available for converting farmers. These were rapidly snapped up as profits from conventional farming dwindled, and the serious expansion (in all sectors except cereals) began. Imports still account for 75% of consumption, but this will drop as more British food becomes available. At the same time, the results of ongoing research into biodiversity have been coming out, reporting quite frightening declines in the populations of all manner of species, especially some of our best-loved birds, but clearly not on organically farmed land. These reports have had a significant effect on the thinking of many people of influence, and have borne out what organic farmers have been saying for decades. So the message really does seem to have got across that organic food is good for you, organic farming is good for the country, and, thanks to the efforts of practising organic farmers and an increasing band of researchers, organic farming works. Marvellous! But what of the future? Are we on a roll? Will all our farmed land become organic? And what will be the influence of the foot and mouth disease (FMD) outbreak?

Firstly, we are not quite on a roll. In fact the rate of conversion has slowed to a trickle, partly because the funding keeps running out, partly because there has been a slight improvement in the fortunes of conventional dairy farming, partly because of (probably) short-term surpluses, particularly of milk, and partly because FMD has put a freeze on much forward planning and consultant activity.

That is how it is from the production end. At the processing and marketing end, there has been a very significant shift in power which is unlikely to benefit British organic farmers. I stated earlier that in the 1980s and early 1990s most organic farms were involved in processing and marketing their produce in some way. Many still are, and a very few have built sizeable businesses, but the big expansion has drawn the big food companies in to exploit the situation. For example, in 1996 over 90% of all of the organic milk processing for the liquid market in Britain was done by organic farmers themselves or companies owned by them; now the situation is reversed and 90% is done by the big four national dairy companies and one American company. The same is true in other sectors, and it is a worrying trend. These companies do not have any particular allegiance either to British agriculture or to organic principles; their function, like that of the big retailers, is to satisfy their shareholders. Their ability to import from all over the world, and their consequent power to drive down farmgate prices, is well known in conventional farming and will be used in the same way in organic.

It is inevitable, therefore, that organic farmers, like their conventional counterparts, face downward pressure on margins, particularly as their job cannot be done properly without significant inputs of labour. So I do not believe that the demand-led expansion will continue at anything like the rate of recent years, which brings us back to government intervention. Here prospects are rosier. There is a general, if slow, shift in EC funding towards agri-environment spending rather than simple production support, and our government is helping to drive this. An organic stewardship scheme which makes ongoing annual payments to farmers of registered organic land is on the cards, and this, if set at a decent level, could be of real benefit in giving farmers the confidence to convert and stay converted. And the post-FMD review, when it comes, seems sure to lead to increased support for local abattoirs and local food links generally, which will benefit organic. So the prettier, livestock-based western side of the country is likely to see ongoing expansion of organic farming. The bigger challenge is to get the arable farmers of the east on board. In spite of organic cereal prices being double those of conventional, very few arable specialists are showing any interest, which means that large quantities of organic cereals for both humans and animals are being imported. Most arable farmers believe that they have gone so far down the chemical road that they cannot come back. This seems a bit disingenuous as the NFU claims that pesticide use has dropped by 23% from its peak a few years ago. This would indicate a movement in the non-chemical direction, so, given the right encouragement, some real progress towards organic could surely be made. Again, a shift from production to stewardship payments would have a significant effect.

In summary, therefore, and in the light of what has happened in the last few years, I must be more optimistic about the future spread of organic farming than I was when writing for the previous edition. But I still feel that, for all the talk of markets and subsidies, the essential ingredient is the quality of the farming. Good organic farming has to be both productive and ecological, and the resulting food has to be excellent. To this end we have to build up our knowledge and understanding of the nature of soils, and what constitutes a really healthy soil from which grow healthy and productive crops and stock. Both existing and incoming organic farmers have to be helped to develop this understanding of the intricacies involved, so that they can bring together their knowledge of manuring, rotations, weather and husbandry to develop really good systems for the land for which they have the privilege of being responsible.

Index